W0082765

defining DARWIN

*Essays on
the History and
Philosophy of*
EVOLUTIONARY
BIOLOGY

defining
DARWIN

MICHAEL RUSE

 **Prometheus Books**

59 John Glenn Drive
Amherst, New York 14228-2119

Published 2009 by Prometheus Books

Defining Darwin: Essays on the History and Philosophy of Evolutionary Biology. Copyright ©
2009 by Michael Ruse. All rights reserved. No part of this publication may be repro-
duced, stored in a retrieval system, or transmitted in any form or by any means, digital,
electronic, mechanical, photocopying, recording, or otherwise, or conveyed via the
Internet or a Web site without prior written permission of the publisher, except in the
case of brief quotations embodied in critical articles and reviews.

Inquiries should be addressed to
Prometheus Books
59 John Glenn Drive
Amherst, New York 14228–2119
VOICE: 716–691–0133, ext. 210
FAX: 716–691–0137
WWW.PROMETHEUSBOOKS.COM

13 12 11 10 09   5 4 3 2 1

Library of Congress Cataloging-in-Publication Data

Ruse, Michael.
    Defining Darwin : essays on the history and philosophy of evolutionary biology /
by Michael Ruse.
        p. cm.
    Includes bibliographical references and index.
    ISBN 978–1–59102–725–6 (hardcover : alk paper)
    1. Evolution (Biology)—History. 2. Natural selection—History. 3. Evolution
(Biology)—Philosophy. 4. Natural selection—Philosophy. 5. Darwin, Charles,
1809–1882. On the origin of the species. I. Title.

QH361.R8736 2009
576.8—dc22

                                                                        2009013240

Printed in the United States on acid-free paper

For:

Matt Day

Fritz Davis

John Kelsay

Joe Travis

CONTENTS

PREFACE 9

ACKNOWLEDGMENTS 11

PART I: DARWIN'S GREAT WORK
One: On the *Origin of Species* 17

PART II: THE EARLY YEARS
Two.   Kant and Evolution 35
Three.   Darwinism and Mechanism 51
Four.   Kicking against the Pricks:
                Alfred Russel Wallace the Rebel 73

PART III: THE MIDDLE YEARS
Five.   Adaptive Landscapes and Dynamic Equilibrium:
                The Spencerian Contribution to Twentieth-
                Century American Evolutionary Biology 95
Six.   Julian Huxley and George Gaylord Simpson
                on Evolution and Ethics 121
Seven.   Evolution and the Novel 153

PART IV: THE LATER YEARS
Eight.   Evo-Devo: A New Evolutionary Paradigm? 177

Nine.   Darwinism Explains Religion (?) 199

Ten.   Evolution as Religion: Are the
                Creationists Right? 215

PERMISSIONS. 243

REFERENCE LIST. 245

INDEX. 257

PREFACE

*T*he child is the father of the man. This holds true as much intellectually as it does emotionally. I trained as a philosopher of science in the 1960s. Always more interested in the fabric of science than in purely logical exercises, I came to evolutionary theory as something worth studying and this has stayed with me all of my life. I read Charles Darwin with as much pleasure today as I did forty years ago. At the same time, I was plunged into the big debate of the day, symbolized by the clash between the Austro-British philosopher Karl Popper and the American physicist-turned-historian Thomas Kuhn. Is science about an objective reality and is the aim to describe and understand that reality as best one can, or is science a far more subjective enterprise, influenced by the culture of the day and as much a creation as an invention? That seemed to me then to be the really important question to be answered, and it seems to me now to be the really important question to be answered. How it is to be answered is also something that comes from this same past. Kuhn told us that to understand science we must turn to its history as well as its present, and as one who was more and more enthused by evolutionary ideas, this was music to my ears. So here I am today: fascinated by evolutionary thinking, believing that the big question is that of objectivity and subjectivity, and convinced that the way to answers is through a study of the science itself, its past and its present. And this is what this collection of essays is all about.

As you will learn very quickly, I do not look upon my work as purely epistemological, that is, to do with truth and justification. I believe that moral issues, within and without science are also vitally important. I would not be interested if this were not so. On the one hand, perhaps more wearing my philosophical hat, it is important for me to understand evolutionary thinking and to defend its standards, its integrity, its results. Today, the forces of darkness, otherwise known as the biblical literalists—in their most modern incarnation, the Intelligent Design Theorists—stand at our gates, trying to enter and to attack evolutionary thinking and its place in school curricula. Fighting these people is itself a moral crusade. On the other hand, perhaps more wearing my historical hat, it is important for me to understand how it is that evolutionary thought has through the ages come to be more than just science, and has played roles in philosophy, in literature, in religion, and more. And always these roles have had underlying social and moral connotations. Finding the truth about these implications and interpretations of evolutionary thought is an important task, and it, too, is moral—both in content and in the importance of being able to separate the real or pure science from all else.

The essays before you—some published, some unpublished, but all tried out in lecture halls—were written in the past ten or so years and have been (usually lightly) revised for this collection. I am pleased that, taken together, they do make a coherent and ongoing whole. They go from the early days of evolutionary thinking down to the present and always the underlying concerns are those detailed in the paragraphs above. I have called the collection *Defining Darwin*, because above all it was Charles Darwin's thinking that made evolutionary ideas worth considering, and because it is the nature of Darwin's own thought and the things that people made of this that motivate me in my inquiries.

ACKNOWLEDGMENTS

*I*t remains only for me to thank the many people over the past decades who have listened to me talk on things evolutionary and who have sharpened my thinking with their penetrating questions. I am much in the debt of my many assistants down through the years: Alan Belk, Christopher Pynes, Jeremy Kirby, Jason Zinser, Charles Alt, Yasha Rohwer, Peter Takacs, Sarah Whylly, and Samantha Muka. As always, Stephen L. Mitchell and the folk at Prometheus have been friends rather than business acquaintances. I am grateful for support from my home institutions—in Canada, the University of Guelph from 1965 to 2000 and then in the US, Florida State University from 2000—and also for research funds from (in Canada) the Social Sciences and Humanities Research Council and then (in the US) the William and Lucyle Werkmeister Fund. Finally, my family—Lizzie, Emily, Oliver, and Edward, not to forget Nigel, Rebecca, and Christian—are what make it all worthwhile. The collection is dedicated to the men with whom I have team taught since I came to Florida. Ultimately, passing on ideas to the next generation is the most important thing that any of us do.

PART I:
DARWIN'S GREAT WORK

*C*harles Robert Darwin (1809–1882) is rightly known as the father of evolutionary theory. He was not the first evolutionist—his grandfather Erasmus Darwin was one of those who came before him—but in his great work, *On the Origin of Species by Means of Natural Selection, or the Preservation of the Favoured Races in the Struggle for Life*, published in 1859, Darwin not only established the fact of evolution beyond reasonable doubt (as they say in the law courts), he proposed natural selection, the mechanism of change that is recognized today as the chief cause behind the development of organisms down through the ages. This first essay of this collection introduces Darwin's great book, sketching the main features within its covers. The ancient Greek philosopher Heraclitus said that you cannot step into the same river twice. It would be a sad reflection on evolutionary studies if today we believed no more about evolution than did Darwin. Indeed, the truth is that there is virtually nothing today in evolutionary studies that corresponds exactly to the facts of the *Origin*. The ancient Greek philosopher Parmenides said that nothing changes, all motion is illusory. This is equally true of the *Origin*. The details are all changed, but the logical structure is the same today as it was for Darwin. As I say in the essay, ultimately Darwin got it right. That is the reason to celebrate him and his theory. We follow in his path.

One

ON THE
ORIGIN OF SPECIES

Mr. Charles Darwin, well known to us all as the author of the delightful travel book *Journal of Researches into the Geology and Natural History of the Various Countries Visited by HMS Beagle* (1839) (better known as *The Voyage of the Beagle*), has now produced an audacious work that will surely be the topic of much conversation for months if not years to come. In his newly published book, *On the Origin of Species by Means of Natural Selection or the Preservation of the Favoured Races in the Struggle for Existence*, to give it the full title, Mr. Darwin openly declares himself a Vestiginarian! Like the unknown author of the *Vestiges of the Natural History of Creation* (1844), Mr. Darwin has embraced transmutationism, what Mr. Herbert Spencer calls "evolution," and what Mr. Darwin himself calls "descent with modification." In his new book—to which we will give the abbreviated title of the *Origin*—Mr. Darwin openly declares that all living creatures (quick and dead) are the end results of a long process of natural development from (as he says) "one or a few forms." And although his book says little about our own species, he makes very clear at the end that we too are part of the picture. "Light will be thrown on the origin of man and of his history." Yes, indeed!

o would open a review of the *Origin* written shortly after the book appeared in 1859. I am writing around the 150th anniversary of the publication, and (declaring an interest) I should tell you that I am the coeditor (with the distinguished historian Robert J. Richards) of the *Cambridge Companion to the "Origin of Species."* So I take it without argument that as a historical work, the *Origin* is of great significance. Goodness, the portrait of Darwin has even replaced Charles Dickens on the back of the English ten-pound note, although rumor has it that the real reason for the shift is that Darwin had a fuller beard than Dickens and thus was less easy to forge! My aim here is to assess the *Origin* from today's perspective, considering its contemporary value as a work of scholarship, and in respects this is a lot less easy to do than from the perspective of, say, 1860. It would be all too easy simply to dismiss it as something irretrievably dated and to leave matters at that. Frankly, I would be worried if one could not dismiss it as dated, for that would suggest that Darwin's ideas have so little interest or value that no one has bothered to try to critique them or to take the discussion further. So remember that here I am not so much suspending critical evaluation as looking at what proved interesting and stimulating in the *Origin*, rather than ways to take cheap shots at the past from the present.

NATURAL SELECTION

Let us turn to the content of Darwin's book; but, as we do so, let us keep in mind something as important today as it was yesterday. Although, when he published the *Origin*, Darwin had been stricken for twenty years with ongoing ill health, he had nevertheless pushed himself to the very first rank of British scientists.[1] His detailed geological studies, based on his five-year-long circumnavigation of the world in the HMS *Beagle* under the captaincy of Robert Fitzroy, were models of empirical inquiry. His massive detailed studies of barnacles, both living and fossil, were exemplary instances of careful and thoughtful study of the world of life. We may or may not agree with Darwin's reasoning. We must respect it and cannot ignore it. Today no less than yesterday we feel at once that we are in the hands of a master. This is not some outsider trying his hand. This is a real professional at work. We can add also that whether or not we agree with Darwin, his warm and easy style makes it exceptionally easy to follow his thinking. Few will come away confused as to the points that he is making. The charming writing of the *Voyage of the Beagle* is once again in evidence. Some may complain—in fact in 1860 his bitter rival, the anatomist Richard Owen, writing in the *Quarterly Review* did complain—

that the style is too easy for a serious work of science. With this I can only disagree. Too often specialists feel that save their writing is opaque and cumbersome, they are not being serious and profound. I am not saying that (in his style) Darwin has the brilliance of Richard Dawkins in the *Selfish Gene*. But he is very good. Would that German metaphysicians and French deconstructivists took lessons from him.

Darwin begins by taking the reader into the world of the breeder—the farmer who wants fatter pigs and hairier sheep, the enthusiast who wants ever-more-fanciful feathers on his pigeon—and he argues that this is a model for change in nature. Here, as throughout his book, Darwin is riding two horses and it is not always clear which back is mounted at any one time. From the one saddle, Darwin thinks that the changes brought about by the breeders are enough to support the changes he supposes through nature. One can almost hear the voice of the Victorian skeptic: "We confess we are not convinced of this. Who ever changes a horse into a cow?" From the other saddle, Darwin prepares the way for his mechanism of change, by showing how selection of the desired is the way that breeders bring about the changes that they effect. This is surely true, although whether selection by breeders is truly analogous to what Darwin calls "natural selection" is indeed the big question at issue. Judging from today, general opinion would be that Darwin was just scratching at the surface, but that he was scratching at the right surface. We now have massive evidence of the power of selection in artificial and semiartificial circumstances, and can change organisms so much that they are really new species, isolated from all others. It is true that we have never changed a horse into a cow, but then neither did nature. They spring from common stock.

Moving to his main mechanism, Darwin first makes reference to the deductions of the Reverend Thomas Robert Malthus,[2] who showed how readily population numbers outstrip supplies of food and space. Here Darwin's genius glows as brightly today as it ever did. With reason, the calculations of Malthus were generally taken to show the impossibility of major change. If you feed the poor from state funds in one generation, you only have more of them in the next. There is bound to be a struggle for existence, save only people practice prudential restraint. Darwin takes the struggle as force that brings on unlimited change, for in the world of animals and plants there can be no prudential restraint, and so in the bloody battle for survival those with advantageous features will tend to succeed. Let us quote Darwin on this:

Let it be borne in mind in what an endless number of strange peculiarities our domestic productions, and, in a lesser degree, those under nature, vary; and how strong the hereditary tendency is. Under domestication, it may be truly said that the whole organisation becomes in some degree plastic. Let it be borne in mind how infinitely complex and close-fitting are the mutual relations of all organic beings to each other and to their physical conditions of life. Can it, then, be thought improbable, seeing that variations useful to man have undoubtedly occurred, that other variations useful in some way to each being in the great and complex battle of life, should sometimes occur in the course of thousands of generations? If such do occur, can we doubt (remembering that many more individuals are born than can possibly survive) that individuals having any advantage, however slight, over others, would have the best chance of surviving and of procreating their kind? On the other hand, we may feel sure that any variation in the least degree injurious would be rigidly destroyed. This preservation of favourable variations and the rejection of injurious variations, I call Natural Selection.[3]

Note that he does not use the alternative phrase "survival of the fittest." This was a phrase coined by Spencer, and Darwin only added it to later editions of the *Origin* at the urging of Alfred Russel Wallace, the codiscoverer of natural selection.

Darwin has many most interesting things to say about natural selection, a process that he argues leads to the wonderful display of fossils in the geological record as well as the multitudinous animals and plants living today. It is still amusing that Darwin appropriates a Christian metaphor when he speaks of history as a "Tree of Life," with us all today at the outer tips of the branches. It is also worth noting in passing that Darwin's mechanism of natural selection shows roots in the Christian faith. It leads to adaptations or contrivances like the hand and the eye. In other words, it supposes that nature is not thrown together randomly but is organized and it works. In the language of the philosophers, the eye shows final cause in seeing and the hand shows final cause in grasping. Here Darwin shows the effects of his training and reading. Passages of the *Origin* would not be out of place in Archdeacon Paley's *Natural Theology*[4] or one of the celebrated *Bridgewater Treatises*,[5] a series of works that appeared in the 1830s intended, in the words of their benefactor, to demonstrate "the power, wisdom, and goodness of God, as manifested in the creation." Darwin would never agree with today's neo-Creationists, the Intelligent Design Theorists, that we need interventions from outside the course of nature to explain the adaptive complexity of organisms, but he does agree with them that adaptation is the chief feature of organic life.

Natural selection leading to adaptation is of course the first of Darwin's big contributions to evolutionary theory. Paradoxically, writing today in 2006 we are more inclined to give him credit than his contemporaries in 1859. Today, although there are still doubters about the ubiquity of natural selection—for instance, the late Stephen Jay Gould[6]— general opinion among professional evolutionists is that Darwin got it right. The chief feature of the organic world is its adaptive or organized complexity (as the late John Maynard Smith used to call it)[7] and natural selection is the only mechanism that speaks to this. (Darwin had a secondary mechanism of sexual selection that is also involved, although some simply subsume it beneath natural selection.) Darwin in the *Origin* does not offer much direct evidence, and today again this is a place where time has moved on. We now have massive evidence of the workings of selection in nature. Very well known are the studies done in the 1950s by H. B. D. Kettlewell[8] on industrial melanism in butterflies, and the studies for the past thirty years by Peter Grant[9] and Peter and Rosemary Grant[10] in the Galapagos Archipelago on Darwin's Finches. But they are just the tip of a huge iceberg. Of great practical importance is the fact that there has never been a medicine introduced that did not become ineffective in its original state because of the rapid, selection-fueled evolution of the attacking microorganisms.

THE CONSILIENCE

The second part of the *Origin*, by far the bigger section, is Darwin's second great contribution to evolutionary theory, and it, too, is as significant today as it ever was. This part is given to a review of biological discussions in different fields in the light of the mechanism of natural selection. The approach taken here, namely, using descent through natural selection to explain phenomena and in turn using the explanations to support descent through natural selection, is not Darwin's own invention. It is the method of scientific argumentation championed by William Whewell in his *Philosophy of the Inductive Sciences*.[11] Whewell referred to this as a "Consilience of Inductions." Just like Newton before him and the geologists establishing plate tectonics after him, Darwin did a terrific and still powerful job of promoting a consilience. Paradoxically and rather sadly, Whewell was so opposed to what he thought were the irreligious aspects of Darwin's theory that, in his position of Master of Trinity College Cambridge, he refused to allow the *Origin* on the college library's shelves.

With superb confidence, which time has only burnished, Darwin takes us through the branches of his subject—instinct, paleontology, the geographical distributions of organisms on the globe, classification, morphology, embryology, and much, much more. It is hard to pick out one topic rather than another, but geography deserves special praise. Why, Darwin asks, are the inhabitants of islands off the coast of Africa similar although not identical to the inhabitants of Africa? Why are they not like the inhabitants of South America? Why conversely are the inhabitants of islands off the coast of South America—including the well-known Galapagos Archipelago—like the inhabitants of South America and not like the inhabitants of Africa? Can there be any explanation except descent with modification?

The most striking and important fact for us in regard to the inhabitants of islands, is their affinity to those of the nearest mainland, without being actually the same species. Numerous instances could be given of this fact. I will give only one, that of the Galapagos Archipelago, situated under the equator, between 500 and 600 miles from the shores of South America. Here almost every product of the land and water bears the unmistakeable stamp of the American continent. There are twenty-six land birds, and twenty-five of those are ranked by Mr Gould as distinct species, supposed to have been created here; yet the close affinity of most of these birds to American species in every character, in their habits, gestures, and tones of voice, was manifest. So it is with the other animals, and with nearly all the plants, as shown by Dr. Hooker in his admirable memoir on the Flora of this archipelago. The naturalist, looking at the inhabitants of these volcanic islands in the Pacific, distant several hundred miles from the continent, yet feels that he is standing on American land. Why should this be so? Why should the species which are supposed to have been created in the Galapagos Archipelago, and nowhere else, bear so plain a stamp of affinity to those created in

America? There is nothing in the conditions of life, in the geological nature of the islands, in their height or climate, or in the proportions in which the several classes are associated together, which resembles closely the conditions of the South American coast: in fact there is a considerable dissimilarity in all these respects. On the other hand, there is a considerable degree of resemblance in the volcanic nature of the soil, in climate, height, and size of the islands, between the Galapagos and Cape de Verde Archipelagos: but what an entire and absolute difference in their inhabitants! The inhabitants of the Cape de Verde Islands are related to those of Africa, like those of the Galapagos to America. I believe this grand fact can receive no sort of explanation on the ordinary view of independent creation; whereas on the view here maintained, it is obvious that the Galapagos Islands would be likely to receive colonists, whether by occasional means of transport or by formerly continuous land, from America; and the Cape de Verde Islands from Africa; and that such colonists would be liable to modifications; the principle of inheritance still betraying their original birthplace.[12]

This explanation is as vital and good today as it was in 1859.

Again, to take a topic much discussed in 1859 by the anatomists and still discussed today by anatomists, why do the forelimbs of animals show similarities—what Richard Owen (1848) called "homologies"[13]—when they are used for different ends? "What can be more curious than that the hand of a man, formed for grasping, that of a mole for digging, the leg of the horse, the paddle of the porpoise, and the wing of the bat, should all be constructed on the same pattern, and should include the same bones, in the same relative positions?" Replies Darwin: "The explanation is manifest on the theory of the natural selection of successive slight modifications, each modification being profitable in some way to the modified form, but often affecting by correlation of growth other parts of the organisation. In changes of this nature, there will be little or no tendency to modify the original pattern, or to transpose parts" (p. 434).[14]

PERSPECTIVE FROM TODAY

There is more, much more, that we could discuss. But the time has now come to turn to discussion of a modern perspective. Darwin established beyond reasonable doubt, as they say in courts of law, that descent with modification accounts for the history of life and the spread of organisms we find today. There is no other way, naturally to account for the facts that Darwin brings to our attention. On the evidence offered in the *Origin*, there is less of a case for the mechanism of natural selection. As has been noted above, time has remedied that omission and today selection rules supreme. What we do not get from the *Origin* (and this was noted by the reviewers of the day) is really helpful information on how features are transmitted from one generation to another and what causes new features. Darwin discusses these issues at length but is more anecdotal than persuasive. As we now know, in the twentieth century the true principles of heredity—going back to the work of the Moravian monk Gregor Mendel living at the same time as Darwin—have been uncovered and developed. We now have a full theory of genetics and it is the backbone of modern evolutionary theory—a backbone supporting the full force of Darwinian selection.

In fairness I should say that there are those who think that some of the advances in the areas of biology explained by selection—paleontology, embryology, and so forth—require new principles that make selection ineffective and perhaps even misleading. Stephen Jay Gould was a big one for arguing in this way, proposing his theory of *punctuated equilibrium* (where evolution proceeds in jumps or starts) as an alternative to Darwinian selection.[15] Unfortunately, neither Gould nor any of the other critics have been able to provide an adequate alternative theory of heredity to explain something like punctuated equilibrium. Mendelism (including its molecular successor) has no place for such jumps or starts.[16] Recently it is the students of embryology (now known as evolutionary development, or evo-devo) who have been trying to break the Darwinian ties. There are all sorts of talk of "constraints" and other such things that make Darwinian adaptive approaches at best otiose and at worst impossible. Gould jumped on this bandwagon too:

> We do need to reformulate, in modern and operational ways, the old notions of organic integrity, and structural determination from the "inside" of genetics and development, thus balancing our former functionalist faith in the full efficacy of adaptationism with positive concepts of internal and structural constraint.[17]

As with punctuated equilibrium, this is to a great extent focusing on non-issues or putting in opposition things no Darwinian would ever dream of denying. It now seems that organisms are not built on new principles each time. They are more like LEGO creations, where the same pieces are used again and again to make very different end results. Why this should be a worry for Darwin or a challenge for natural selection entirely escapes me. Of course it means that you cannot do everything. As in LEGO, you are constrained by the pieces that you have. But whoever thought you could do everything? In LEGO, it is simply amazing how much you can do, and the same is true in the biological world.

Finally, let us ask the question that (in America particularly) is as much at the fore of the minds of people today as it was at the fore of the minds of Darwin's contemporaries. How far can one go with Darwin if one is a practicing, believing Christian? If one believes that Jesus was God incarnate, who died on the cross for our sins, and makes possible our eternal salvation, can one accept descent with modification? My answer today is that which most people, including most practicing Christians (excluding the American South) gave back then. It is many years now since either scientists or the layman felt constrained by literal readings of Genesis. No one believes in the date set by Archbishop Ussher, that the world is but six thousand years old. More positively, remember how Darwin's mechanism of natural selection speaks directly to the significance of design, of final cause. If evolution be true, there is no reason why God should not have created through law rather than miracle. After all, would we say that the Briton who builds a power loom to make his cloth is less worthy as a man than the one who persists in using a hand loom?

Christians in the past and Christians today have reservations about *Homo sapiens*. Darwin clearly wants to include us in the picture, and as far as our mortal frame is concerned, there can be no objection. But as creatures made in the image of God, as beings with immortal souls, many believers will balk. They will call for divine intervention. And perhaps at this point Darwin and his followers can lean back and give Christians their case. A soul may be real—more real than most—yet not a proper object of scientific investigation. I am not saying that one must believe in souls—Darwin did not and neither do I—but I am saying that if Christians want to believe in souls, who is to deny them that pleasure?

How does one judge the *Origin* overall from 2006? I find a comparison with Freud very enlightening. (A comparison with Marx may make the same point.) I have read a great deal of Freud's writings very carefully. I do not think of myself as a Freud expert as (immodestly) I would think of myself as a Darwin expert, but I think I can talk knowledgeably about his ideas. Some, if not just about all, of Freud's writings I find very exciting. He, like Darwin, can write in a terrific way and make his ideas understandable and compelling at the moment. I recognize fully that Freud has been very influential. Has he been more influential than Darwin? I don't know and don't much care. They have both been very influential, and not just in their own fields but in culture generally. There are many ideas in Freud that I think are absolutely spot on—much more so than the ideas of some of his followers. For instance, it seems to me to be absolutely the case that homosexual males tend to have close relationships with their mothers and difficult relationships with their fathers. Not all homosexuals, not all mothers, not all fathers, but many—more so than with heterosexual males.[18]

Yet when all is said and done I just don't think that Freud got it. I just don't think that he had a basic theory of human nature that works or that is true of the empirical world. Insights, but not truth. In the case of homosexual males, Freud argues that it is the dysfunctional triangle that brings on the orientation: You cannot successfully resolve the Oedipal tensions because you want to have sex with mother, you know you can't, and so you revert to a childhood state of being attracted to males, and the tension is resolved. That seems to me just plain wrong. Freud got it backwards. Boys are born with their orientations and it is they who elicit the behavior rather than the parents who impose it. A gay son in the making wants to play with mum a lot more than with dad, and so the attitudes evolve and harden. There was no chap called Oedipus, it was not engrained in human nature by Lamarckian (inheritance of acquired characteristics) processes, and I very much doubt that most male homosexuals really want to romp between the sheets with their mothers.

Darwin is different. You can criticize as much as you like. He did get things wrong and he did offer too little evidence too often. But ultimately he was right. He grasped the truth about the way that the world is. It is as simple as that. Evolution is true and natural selection is its mechanism. No more, but certainly no less. So let us end by going back to the *Origin* and to its final words. We are given a paragraph of poetry in prose, somewhat ironically modified from a review Darwin read in 1838, by the Scottish physicist David Brewster, on a work of positivism by August Comte.[19] You will not be surprised to learn that Brewster was praising God. The naturalist Charles Darwin praises evolution:

> It is interesting to contemplate an entangled bank, clothed with many plants of many kinds, with birds singing on the bushes, with various insects flitting about, and with worms crawling through the damp earth, and to reflect that these elaborately constructed forms, so different from each other, and dependent on each other in so complex a manner, have all been produced by laws acting around us. These laws, taken in the largest sense, being Growth with Reproduction; Inheritance which is almost implied by reproduction; Variability from the indirect and direct action of the external conditions of life, and from use and disuse; a Ratio of Increase so high as to lead to a Struggle for Life, and as a consequence to Natural Selection, entailing Divergence of Character and the Extinction of less-improved forms. Thus, from the war of nature, from famine and death, the most exalted object which we are capable of conceiving, namely, the production of the higher animals, directly follows. There is grandeur in this view of life, with its several powers, having been originally breathed into a few forms or into one; and that, whilst this planet has gone cycling on according to the fixed law of gravity, from so simple a beginning endless forms most beautiful and most wonderful have been, and are being, evolved.[20]

NOTES

1. J. Browne, *Charles Darwin: Voyaging. Volume I of a Biography* (New York: Knopf, 1995); J. Browne, *Charles Darwin: The Power of Place. Volume II of a Biography* (New York: Knopf, 2002).

2. T. R. Malthus, *An Essay on the Principle of Population*, 6th ed. (London: Everyman, 1914 [1826]).

3. C. Darwin, *On the Origin of Species by Means of Natural Selection, or the Preservation of Favoured Races in the Struggle for Life* (London: John Murray, 1859), pp. 80–81.

4. W. Paley, *Natural Theology* (Collected Works: IV) (London: Rivington, 1819 [1802]).

5. C. C. Gillespie, *Genesis and Geology* (Cambridge, MA: Harvard University Press, 1950).

6. S. J. Gould, *The Structure of Evolutionary Theory* (Cambridge, MA: Harvard University Press, 2002).

7. J. Maynard Smith, "The Status of Neo-Darwinism," in C. H. Waddington, ed., *Towards a Theoretical Biology* (Edinburgh: Edinburgh University Press, 1969).

8. H. B. D. Kettlewell, *The Evolution of Melanism* (Oxford: Clarendon, 1973).

9. P. R. Grant, *Ecology and Evolution of Darwin's Finches* (Princeton, NJ: Princeton University Press, 1986).

10. R. B. Grant and P. R. Grant, *Evolutionary Dynamics of a Natural Population: The Large Cactus Finch of the Galapagos* (Chicago: University of Chicago Press, 1989).

11. W. Whewell *The Philosophy of the Inductive Sciences,* 2 vols. (London: Parker, 1840).

12. Darwin, *On the Origin of Species*, pp. 397–98.

13. R. Owen, *On the Archetype and Homologies of the Vertebrate Skeleton* (London: Voorst, 1848).

14. Darwin, *On the Origin of Species*, p. 434.

15. N. Eldredge and S. J. Gould, "Punctuated Equilibria: An Alternative to Phyletic Gradualism," in T. J. M. Schopf, ed., *Models in Paleobiology* (San Francisco: Freeman, Cooper, 1972), pp. 82–115.

16. G. L. Stebbins and F. J. Ayala, "Is a New Evolutionary Synthesis Necessary?" *Science* 21 (1981): 967–71.

17. Gould, *Structure of Evolutionary Theory*, p. 1057.

18. S. Freud, "Three Essays on the Theory of Sexuality," in *The Standard Edition of the Complete Psychological Works of Sigmund Freud*, vol. 7 (London: Hogarth, 1955 [1905]), pp. 125–243.

19. This is the passage by Brewster:

In considering our own globe as having its origin in a gaseous zone, thrown off by the rapidity of the solar rotation, and as consolidated by cooling from the chaos of its elements, we confirm rather than oppose the Mosaic cosmogony, whether allegorically or literally interpreted . . . .

In the grandeur and universality of these views, we forget the insignificant beings which occupy and disturb the planetary domains. Life in all its forms, in all its restlessness, and in all its pageantry, disappears in the magnitude and remoteness of the perspective. The excited mind sees only the gorgeous fabric of the universe, recognizes only its Divine architect, and ponders but on its cycles of glory and desolation. (D. Brewster, "Review of Comte, *Cours de Philosophie Positive*," *Edinburgh Review* 67 [1838]: 301)

Consider now the closing words of Darwin's first outline of his theory (commonly known as the *Sketch* of 1842), repeated with little change in the *Origin*. I italicize words echoing Brewster:

There is a simple *grandeur* in the *view* of *life* with its powers of growth, assimilation and reproduction, being *originally* breathed into matter under one or a few *forms*, and that whilst this our *planet* has gone circling on according to fixed laws, and land and water, in a *cycle* of *change*, have gone on replacing each other, that from so simple an *origin*, through the process of gradual selection of infinitesimal changes, endless *forms* most beautiful and most wonderful have been evolved. (C. Darwin and A. R. Wallace, *Evolution by Natural Selection* [Cambridge: Cambridge University Press], p. 87)

20. Darwin, *On the Origin of Species*, pp. 489–90.

PART II:
THE EARLY YEARS

*9*n culture as in biology, things do not spring into existence fully formed and mature. Ideas, scientific theories have origins, and this is certainly true of Darwinian evolutionary theory. It would be easiest just to talk about predecessors—Erasmus Darwin; the French biologist Jean Baptiste de Lamarck; the Scottish publisher and author of *Vestiges* Robert Chambers. But it is often more interesting and rewarding to talk about those who *reject* ideas. The natural inclination is to say that these people were history's losers, the people who did not show the way forward. There is often truth in this, but not necessarily. Opponents sometimes have very good reasons for their opposition, and when finally the breakthroughs do come, we can see the importance of these new moves precisely because they do speak to the earlier objections. This is certainly true of the great German philosopher Immanuel Kant, who tried—really tried—to accept evolutionary ideas, but in the end just could not because he could not see how blind law can lead to the forward-looking complexity, the teleology of the living world. Kant was one of those who posed the big problem, one taken up and reaffirmed by the French comparative anatomist Georges Cuvier, that had to be answered by Darwin, something he did with his mechanism of natural selection. Kant was wrong, but he was importantly wrong.

The second essay in this section looks in more detail at Darwin's hypothesis of natural selection, trying to understand its true nature and the ways in which it functioned trying to speak to the sorts of problems that Kant raised. Like many philosophers, particularly those of my generation, I am fascinated by the use of metaphor in science, and in this essay I explore the metaphor of nature as machine, looking specifically at Darwin's thinking in this regard. Distinguishing two senses of the metaphor, I show that in some respects Darwin embraces the metaphor fully but that in other respects he is perhaps a little more guarded. In the more guarded respects—respects that we have abandoned today—Darwin perhaps shows his links to his preevolutionary past. This would be no surprise. In the ways that he does embrace the metaphor, he shows both how he speaks to the problems posed by earlier thinkers—Kant particularly—and how he overcomes them in what is now a thoroughly modern way. Darwin was no atheist, nor even at the time of writing the *Origin* an agnostic. (He was a deist, believing in God as an unmoved mover.) But certainly a subtext of this essay is that of showing how religious ideas were constantly retreating before the successes of science.

The final essay of this section turns to Alfred Russel Wallace, the codiscoverer of natural selection. Wallace is a sympathetic character and one needs to beware of oversympathizing with one who has always had to play second fiddle to Charles Darwin. My aim is less with defending (or attacking) Wallace and more with showing how, for all of his brilliance, he represented a way of doing science that was fading fast at the time of the *Origin* and in the decades just following. Wallace wanted to invoke spirit forces to explain human evolution, and that was just not on. It was not so much that he was wrong, but that that was no longer the way that Darwin and his friends (notably Thomas Henry Huxley) did science. We see therefore that in the history of science, in the history of evolutionary theorizing, it was not just a question of new ideas, new facts, new metaphors (or old metaphors extended)—it was also a question of how one does science itself and how religious and quasi-religious ideas are simply no longer allowed in science. This incidentally is the reason why today's scientists object so strongly to Creationist ideas like so-called Intelligent Design Theory, something that supposes an intelligence produces adaptive complexity. Even if selection could not do the job, and evolutionists today think that it can, appealing to intelligences (especially supernatural intelligences) is just not the way one does science.

Two

KANT AND EVOLUTION

Where did Immanuel Kant stand on the question of organic evolution? This has been a topic of much debate, and it seems that it is still ongoing. Some think that he was close if not committed to evolution.[1] This would fit with a general dynamic view of nature, and harmonizes nicely with Kant's hypothesis about a natural origin of the universe—his formulation of what is known as the "nebular hypothesis." Others think that he was not in fact an evolutionist, but that this was a contingent matter. Kant could have been an evolutionist; it was just that he did not think the facts were favorable to the idea.[2] And yet others—and I am one—think that Kant was not an evolutionist and that his opposition was deep and theoretical. Given his philosophy of nature, he simply could not have been an evolutionist.[3]

In major part, I think the confusion about Kant's position represents confusion in Kant's own thinking. To say something that a historian of philosophy can say but that a historian of science would never say, Kant simply was not dealing with a full deck and so he could not get things right. He did not know about the mechanism of natural selection, the mechanism that gives an adequate law-bound answer to the issue of organic origins. Hence, Kant was groping with the problems without the tools to give the full answers. However, I come now to praise Kant, not to bury him. Apart from the intrinsic interest of whatever Kant has to say on a topic—and although I think his position incomplete and inadequate, I think there are major insights of relevance to our understanding of organisms today—there is also some historical interest in trying to disentangle Kant's thinking. Kant's thinking was influential on others, specifically the German-trained, French comparative anatomist Georges Cuvier, the greatest of all opponents of evolutionary thought. If the interpretation of Kant that I endorse is right, then this goes a long way to explaining the opposition of Cuvier, which was also more theoretical than empirical. And since Charles Darwin took Cuvier's concerns very seriously, accepting Cuvier's problem-situation if denying Cuvier's nihilism, Kant's thought can then plausibly be said to be a factor in the formulation of the causal theory of evolution that almost all biologists endorse today. One can go one step earlier than the French historian and philosopher of science, Michel Foucault, who wrote, "Seen in its archeological depth, and not at the more visible level of discoveries, discussion, theories, or philosophical options, Cuvier's work dominates from afar what was to be the future of biology." More fully: "Seen in its archeological depth, and not at the more visible level of discoveries, discussion, theories, or philosophical options, Kant's work dominates from afar what was to be the future of biology."[4]

KANT ON ORGANISMS

Start with Kant on organisms, whose crucial thinking on the subject comes in the second part of the *Critique of Judgement*. In the tradition that goes back to Aristotle, Kant saw that the significant feature of organisms is that they are organized—they and their parts seem as if fashioned for the ends of survival and reproduction. In other words, organisms demand explanation in terms of final causes as well as efficient and other causes. Yet, Kant saw that final causes are problematic. From the seventeenth century, many scientists—particularly scientists in the physical sciences—had tried to expel final causes as unneeded, confusing, and unacceptably theological. Francis Bacon, for instance, had wittily likened them to vestal virgins—beautiful but sterile. In line with this kind of thinking, Kant was uncomfortable with final-cause talk, because it does seem to imply design, and this is simply not acceptable in science. We are only allowed to talk in terms of material or mechanical causes. "Hence if we supplement natural science by introducing the conception of God into its context for the purpose of rendering the finality of nature explicable, and if, having done so, we turn round and use this finality for the purpose of proving that there is a God, then both natural science and theology are deprived of all intrinsic substantiality." Kant was unbending on this. "This deceptive crossing and recrossing from one side to the other involves both in uncertainty, because their boundaries are thus allowed to overlap."[5]

But—and here Kant fitted well in line with those who worked on organisms—Kant recognized that we simply cannot do without final-cause thinking. Heuristically, in biology teleology is absolutely essential. We need the maxim "an organized natural product is one in which every part is reciprocally both end and means." We simply cannot do biology without assuming final cause. "It is common knowledge that scientists who dissect plants and animals, seeking to investigate their structure and to see into the reasons why and the end for which they are provided with such and such parts, why the parts have such and such a position and interconnexion, and why the internal form is precisely what it is, adopt the above maxim as absolutely necessary."[6] Scientists cannot do biology in any other way. Teleological thinking is not a luxury; it is a necessity. Life scientists

> say that nothing in such forms of life is in vain, and they put the maxim on the same footing of validity as the fundamental principle of all natural science, that nothing happens by chance. They are, in fact, quite as unable to free themselves from this teleological principle as from that of general physical science. For just as the abandonment of the latter would leave them without any experience at all, so the abandonment of the former would leave them with no clue to assist their observation of a type of natural things that have once come to be thought under the conception of physical ends.[7]

So how are we to solve the problem, what Kant called an "antimony," of needing to use final-cause talk and yet recognizing that only material-cause talk is acceptable in physical science or any science that claims to be talking of objective reality? Here the Kantian metaphysics comes into play—phenomenally we can see no design in nature, but noumenally it is possible that there is design. God may be standing behind everything, but this is for things in themselves and not for the phenomenal world as we know it.

> It is at least possible to regard the material world as a mere phenomenon, and to think something which is not a phenomenon, namely a thing-in-itself, as its substrate. And this we may rest upon a corresponding intellectual intuition, albeit it is not the intuition we possess. In this way, a supersensible real ground, although for us unknowable, would be procured for nature, and for the nature of which we ourselves form part. Everything, therefore, which is necessary in this nature as an object of sense we should estimate according to mechanical laws. But the accord and unity of the particular laws and of their resulting subordinate forms, which we must deem contingent in respect of mechanical laws—these things which exist in nature as an object of reason, and, indeed, nature in its entirety as a system, we should also consider in the light of teleological laws. Thus we should estimate nature on two kinds of principles. The mechanical mode of explanation would not be excluded by the teleological as if the two principles contradicted one another.[8]

We may (must) suppose God, but we cannot prove it. "All that is permissible for us men is the narrow formula: We cannot conceive or render intelligible to ourselves the finality that must be introduced as the basis even of our knowledge of the intrinsic possibility of many natural things, except by representing it, and, in general, the world, as the product of an intelligent cause—in short, of a God."[9]

For Kant then, teleological thinking is a regulative principle; it is a necessary heuristic. It is not a condition of rational thinking in the way that the mechanical philosophy is. We cannot think of the world except as causally, for instance. We can certainly look at organisms without thinking of final causes. But as soon as we start to study them, to understand them, final-cause thinking comes into play—has to come into play. Final causes are part of the filter, the lens, through which we study the world. They are our doing: similar to things like causality in that we impute them to the world, but less strong than causality because we can think without them even though we cannot work without them. They are regulative. "Strictly speaking, we do not observe the *ends* in nature as designed. We only read this conception *into* the facts as a guide to judgement in its reflection upon the products of nature. Hence these ends are not given to us by the Object."[10]

KANT ON EVOLUTION

Now how does all of this relate to organic evolution? As I have said above, I argue that Kant was not just an opponent of such evolution, but was theoretically deeply against the idea. This follows precisely because of his commitment to adaptation and final cause. As I have noted, you might be surprised at this. Particularly given his interest in the nebular hypothesis, you might think that Kant would have been at least empathetic to evolutionary speculations. However, far from Kant having seen evolution as being in some sense the answer to the nature of organisms, perhaps throwing light on final causes, he rather saw (organic) evolution as an impossibility. Specifically, he saw final causes—that is to say, situations where one feels obligated to invoke an end-directed kind of understanding—as being a barrier to true evolutionary thought:

> It is, I mean, quite certain that we can never get a sufficient knowledge of organized beings and their inner possibility, much less get an explanation of them, by looking merely to mechanical principles of nature. Indeed, so certain is it, that we may confidently assert that it is absurd for men even to entertain any thought of so doing or to hope that maybe another Newton may some day arise, to make intelligible to us even the genesis of but a blade of grass from natural laws that no design has ordered. Such insight we must absolutely deny to mankind.[11]

This does not mean that there cannot objectively be final cause without design (without a Designer, that is), for this would presume to know what we cannot know. What it does mean is that we cannot ourselves, at the phenomenal level, expect to get a mechanical explanation.

Kant did not think that the idea of organic evolution is silly. Indeed, like some of the evolutionists of his day—for instance, Erasmus Darwin (whose specific thinking Kant could hardly have known at the time of writing the *Critique of Judgement*)—Kant thought that the isomorphisms between the parts of different organisms rather points in the way of evolution. "This analogy of forms, which in all their differences seem to be produced in accordance with a common type, strengthens the suspicion that they have an actual kinship due to descent from a common parent." Kant even indeed went on to spell things out, speaking of our ability to "trace in the gradual approximation of one animal species to another, from that in which the principle of ends seems best authenticated, namely from man, back to the polyp, and from this back even to mosses and lichens, and finally to the least perceivable stage of nature."[12] But ultimately it appears that these connections are all ideal—connections in theory and not in actuality. There is no common descent. Evolution is untrue.

Why? In one way, let me agree that Kant did rather suggest that it is a contingent matter that evolution is false. "An hypothesis of this kind may be called a daring venture on the part of reason; and there are probably few even among the most acute scientists to whose minds it has not sometimes occurred. For it cannot be said to be absurd, like the *generatio aequivoca*, which means the generation of an organized being from inorganic matter." In Kant's own language, there is nothing a priori self-contradictory about the idea of organic evolution. It is certainly logically possible that animals move from the water to the marshes, and thence to the land, changing and adapting as they go along. Such a notion of evolution is not like a round square that never could exist, even in principle. Nevertheless, the facts of nature go against evolution. "Experience offers no example of it. On the contrary, as far as experience goes, all generation known to us is *generatio homonyma*. It is not merely *univoca* in contradistinction to generation from an unorganized substance, but it brings forth a product which in its very organization is of like kind with that which produced it, and a *generatio heteronyma* is not met with anywhere with the range of our experience.[13]

But then Kant made it clear that the trouble with evolution is more than something just contingent. Evolution may not be a logical impossibility, but it goes against the final-cause thinking we use in biology. Organisms are organized in the way that they are, and moving across from one species to another would disrupt this organization in a way fatal to the intermediaries. "For in the complete inner finality of an organized being, the generation of its like is intimately associated with the condition that nothing shall be taken up into the generative force which does not also belong, in such a system of ends, to one of its undeveloped native capacities." Breaking from this inner finality would be too disruptive. Assuredly, "the principle of teleology, that nothing in an organized being which is preserved in the propagation of the species should be estimated as devoid of finality, would be made very unreliable and could only hold good for the parent stock, to which our knowledge does not go back."[14] Kant was not the clearest of writers and these passages are no exception. But the basic idea is clear: it is evolution or final cause, and final cause wins. Contrivance is a fact of nature requiring final-cause understanding—the complexity of nature demands that we think in terms of ends appropriate to an intelligence. Evolution is a blind-law explanation, precluding final-cause understanding—no blind-law explanation can yield phenomena requiring understanding in terms of ends appropriate to an intelligence. Hence, evolution cannot in principle explain contrivance, and so must be false. (Note that in speaking of "ends appropriate to an intelligence," this is not as such to imply that there is an intelligence. This is the whole point of Kant's intellectual wriggling.)

OPUS POSTUMUM

What about the last, unfinished work, the *Opus Postumum*? (This was begun in 1796, the year Kant retired from teaching. Kant died in 1804, although his mental powers had faded in his final years.) It is here that one can make the strongest case for Kant's evolutionism, and let me say that it is a good case. For a start, we find here that by now Kant knew of Erasmus Darwin's evolutionary work, *Zoonomia*. (This was published between 1794 and 1796. It was immediately translated into German, with the first two parts appearing in 1795 and the rest by 1799.) More significantly, we have several passages (by Kant) that look very close to endorsing some form of transmutationism:

> The organized creatures form on earth a whole according to purposes which [can be thought] *a priori*, as sprung from a single seed (like an incubated egg), with mutual need for one another, preserving its species and the species that are born from it.

Also, revolutions of nature which brought forth new species (of which man is one.)[15]

Not only does the vegetable kingdom exist for the sake of the animal kingdom (and its increase and diversification), but men, as rational beings, exist for the sake of others of a different species (race). The latter stand at a higher level of humanity, either simultaneously (as, for instance, Americans and Europeans) or sequentially. For instance, if our globe (having once been dissolved into chaos, but now being organized and regenerating) were to bring forth, by revolutions of the earth, differently organized creatures, which, in turn, gave place to others after their destruction, organic nature could be conceived in terms of a sequence of different world-epochs, reproducing themselves in different forms, and our earth as an organically formed body—not one formed merely mechanically.

How many such revolutions (including, certainly, many ancient organic beings, no longer alive on the surface of the earth) preceded the existence of man, and how many (accompanying, perhaps, a more perfect organization) are still in prospect, is hidden from our inquiring gaze—for, according to Camper, not a single example of a human being is to be found in the depth of the earth.[16]

Nature organizes matter in manifold fashion—not just by kind, but also by stages. Not to be comprehended: That there are to be discovered in the strata of the earth and in mountains, examples of the former kinds of animals and plants (now extinct)—proofs of previous (now alien) products of our living, fertile globe. That its organizing force has so arranged for one another the totality of plants and animals, that they, together, as members of a chain, not merely in respect of their nominal character (similarity) but their real character (causality) which points in the direction of a world organization (to unknown ends) of the galaxy itself.[17]

Let me make three comments about these and related passages. First, they do not as such affect the assessment I have made of the biological thinking of the *Critique of Judgement*. The *Opus Postumum* comes in the decade after this work. Kant could have changed his mind, for better or for worse. He would not have been the first philosopher to have done so.

Second, even if they represent a complete about-face for Kant, they do not affect the subsequent course of history. The *Opus Postumum* went unpublished for many decades, and only in the twentieth century did good editions start to appear (in English, but ten years ago). Historically, the work and its contents are irrelevant.

Third, I am not at all convinced that they do represent an about-face. I am certainly not convinced that Kant now becomes an evolutionist. A number of people read *Zoonomia* almost as soon as it was translated into German, but the mechanical-materialism of that work simply did not convince. No one was picking up on and endorsing the evolutionism. (Richards makes this point about Schelling.)[18] Kant's reference to *Zoonomia* seems to be about the teleological nature of organisms, not their evolution. Apparently the Englishman stimulated Kant to think of "organic bodies, every part of which is there for the sake of the other, and whose existence can *only* be thought in a system of purposes (which must have an immaterial cause."[19] And the passages quoted above, although perhaps implying and endorsing an upward progress to life through time, do not make this an actual evolution. For others at the time—Schelling, Goethe, Hegel (soon after)—one had the upwards thrust of life, with ever-greater manifestations of basic forms (*Baupläne*, or archetypes), but it is more an ideal pattern (perhaps seeds springing forth) than an actual physical evolution:

> Nature is to be regarded as a *system of stages*, one arising necessarily from the other and being the proximate truth of the stage from which it results: but it is not generated *naturally* out of the other but only in the inner Idea which constitutes the ground of Nature. *Metamorphosis* pertains only to the Notion as such, since only its alteration is development. But in Nature, the Notion is partly only something inward, partly existent only as a living individual: *existent* metamorphosis, therefore, is limited to this individual alone.[20]

Of course, you might say, none of this is very satisfactory. That is precisely my point. It is not very satisfactory. Kant and his fellows did not have natural selection. The issue of importance here is that one cannot consider even the aged Kant to have been an evolutionist, and—even though his thinking was incomplete because of his ignorance of the needed science—it is not to deny the importance of Kant historically or the value of his philosophy of biology, especially as expressed in the *Critique of Judgement.*

GEORGES CUVIER

Why do I claim that there is much for us today to learn from Kant? Particularly important is the emphasis on final-cause-type thinking—its necessity whatever its status—and Kant's penetrating analysis of its nature. In fact, he anticipates (an anticipation that I have never seen acknowledged) one of the most popular contemporary analyses of such teleological thinking. The philosopher Larry Wright has argued that when we say that Z is the function of X—in traditional language, when we say that Z is the final cause of X—we are not only saying that X is there because it does Z, we are also saying that Z is (or happens as) a result or consequence of X's being there. So, in terms of familiar examples: "Not only is chlorophyll in plants *because* it allows them to perform photosynthesis, photosynthesis is a *consequence* of the chlorophyll's being there. Not only is the valve-adjusting screw there *because* it allows the clearance to be easily adjusted, the possibility of easy adjustment is a *consequence* of the screw's being there."[21] Notice that all of this has to be understood in a generic sense, otherwise we run into missing-goal-object problems and the like. X does not always have to do Z in every case—it may be that X does Z on only a few occasions—but it must do it sometimes, and these must matter.

I need hardly say that this is pure and undiluted Kant, for like the great German philosopher, Wright is offering a proposal that involves a two-way causal connection. "A does B. A exists because it does B." In Kant's language: "I would say that a thing exists as physical end *if it is* (though in a double sense) *both cause and effect of itself.*"[22] In Wright's language: "When we give a functional explanation of X by appeal to Z ("X does Z"), Z is always a consequence or result of X's being there (in the sense of "is there" sketched above). So when we say that Z is the function of X, we are not only saying that X is there because it does Z, we are also saying that Z is (or happens as) a result or consequence of X's being there." Of course, what Wright knows and Kant did not know is that natural selection shows how something can be both cause and effect of itself—the success of the adaptation leads to its continuance, which in turn leads to the adaptation being used again. The hand and the eye lead to survival and reproduction, leading to more hands and eyes. But it is Kant who has captured the logic of the situation—a situation that we do not find in the physical sciences. And of course, as Kant saw, the reason why we have this kind of peculiar situation in the life sciences is because organisms seem as if designed—they lead to their replication because of their distinctive natures. So even though we may judge Kant's analysis incomplete—as I would—we can nevertheless use his thinking for our own understanding of the logic of biological explanation—as I would.

But as I said above, more immediately pressing is the influence that Kant may have had on others—others who played a key role in the history of the idea of evolution. The opposition that Kant has to evolution—you cannot square it with teleology—is precisely the opposition that is expressed by Georges Cuvier. For the German-trained Cuvier, the key to understanding the organism lies in the fact that it is not simply subject to the physical laws of nature, but that it is organized, with the parts directed to the end of the functioning whole—each individual feature playing its role in the overall, end-directed scheme of things. This Cuvier referred to as the "conditions of existence," writing as follows:

> Natural history nevertheless has a rational principle that is exclusive to it and which it employs with great advantage on many occasions; it is the *conditions of existence* or, popularly, *final causes*. As nothing may exist which does not include the conditions which made its existence possible, the different parts of each creature must be coordinated in such a way as to make possible the whole organism, not only in itself but in its relationship to those which surround it, and the analysis of these conditions often leads to general laws as well founded as those of calculation or experiment.[23]

But how now is one to translate the teleology into a practical working science? Here we move to Cuvier's field of scientific interest and expertise, (animal) morphology or anatomy. He made detailed studies of animal after animal, thinking that the conditions of existence yielded a working guide for the investigator. This corollary, as we might call it, was referred to by Cuvier as the "correlation of parts." He argued that in order to be an integrated functioning being, every part of an organism had to be slotted in harmoniously with every other part. "It is in this mutual dependence of the functions and the aid which they reciprocally lend one another that are founded the laws which determine the relations of their organs and which possess a necessity equal to that of metaphysical or mathematical laws." And then, linking this back to the conditions of existence (of which the correlation of parts is really just a physical manifestation), "it is evident that the seemly harmony between organs which interact is a necessary condition of existence of the creature to which they belong and that if one of these functions were modified in a manner incompatible with the modifications of the others the creature could no longer continue to exist."[24]

The point is that the correlation of parts supposedly gave the anatomist a tool to make predictions. Pretend, for instance, you have only the tooth of an animal. From its design you can tell that it is for meat-eating rather than for chewing vegetable matter. Then, you work outwards. If the owner is a carnivore, there is little point in its having hooves like a deer or a stomach like a cow or the armor of a tortoise or any of the other attributes that one associates with vegetarian prey. Rather, one needs claws and agility and intelligence and so forth. Thus you can begin to infer what the whole animal must look like. And, one should say, Cuvier felt that this line of argumentation showed its worth again and again, for on several occasions having been given but a fossil fragment of an unknown beast, he was able to infer the whole form—a prediction triumphantly confirmed when later the whole animal was discovered.

The conditions of existence yield a second corollary, the "subordination of characters," and it was through this that Cuvier thought he was able to bring order to the animal world, dividing it into four basic groups or *embranchements*.

> For a good classification . . . we employ an assiduous comparison of creatures directed by the principle of the subordination of characters, which itself derives from the conditions of existence. The parts of an animal possessing a mutual fitness, there are some traits of them which exclude others and there are some which require others; when we know such and such traits of an animal we may calculate those which are

coexistent with them and those which are incompatible; the parts, properties, or consistent traits which have the greatest number of these incompatible or coexistent relations with other animals, in other words, which exercise the most marked influence on the creature, we call *caracteres importants*, *caracteres dominateurs*; the others are the *caracteres subordones*, and there are thus different degrees of them.[25]

Suppose you have in place a backbone. Then you know that many features to be found in the animal world no longer are possible for this particular animal. It must now have characters and only those characters that are part and parcel of being a vertebrate. Suppose now you have in place the backbone of a whale. It is no longer possible to have the limbs of a land predator like a tiger, nor the teeth of such an animal, nor its stomach or its brain, or many other things. Once you have gone the route of sea mammal, then you are constrained in many ways except for one outlet only. This gives rise to the proper way to classify—being ever more restricted in the choices open to the production of a functioning organism. And so we get the four basic groups (*embranchements*) of animals: the vertebrates, the molluscs, the articulates, and the radiates. "I have found that there exist four principal forms, four general plans, upon which all of the animals seem to have been modelled and whose lesser division, no matter what names naturalists have dignified them with, are only modifications superficially founded on development or on the addition of certain parts, but which in no way change the essence of the plan."[26]

AGAINST EVOLUTION

This then is Cuvier's brilliant teleological program for the animal world, very much in the spirit of Kant. But there is more than this, for Cuvier was right in line with Kant's denial that final cause and evolution can be held be held consistently by one and the same person. Like the philosopher, at one level Cuvier was happy to appeal to the empirical evidence. Napoleon had gone campaigning in Egypt and—taking his savants with him—mummified humans and animals had been returned to France. But although these are very old, there is no trace of evolutionary change. The

Egyptians "have not only left us representations of animals, but even their identical bodies embalmed and preserved in the catacombs."[27] If not evolutionary, then what is the nature of organic origins? Cuvier admitted he did not have much positive idea. Beyond allowing that there seems to be a roughly progressive history to life (as revealed in the fossil record, which he himself was doing much to open up), Cuvier had little to say. "I do not pretend that a new creation was required for calling our present races of animals into existence. I only urge that they did not occupy the same places, and that they must have come from some other part of the globe."[28]

Ultimately, however, in the steps of Kant, it was final cause that spelt the doom of evolution. With the German philosopher, Cuvier just could not see how you could get change across a species barrier without disruption of the organism too great to be borne. The intricacies of adaptation exist specifically to aid the organism in the life and role that it occupies. Hybrids are neither fish nor fowl—adapted neither for water nor for air—and as such cannot possibly survive and reproduce. To use a mathematical analogy, the sum of the internal angles of an n-sided polygon equals 2n-4 right angles. You go, to take the example of the move from triangle to quadrilateral, from 2 right angles to 4 right angles. You simply cannot have a plane-sided figure whose internal angles add up to 3 right angles. That is simply an impossibility. And so also is an organism between two species, and especially an organism between two *embranchements*.

Speaking parenthetically, the mathematical analogy ought to have appealed to Kant, although actually the impossibility of a polygon with internal angles summing to 2n-2 right angles would seem to be synthetic a priori, as are the other truths of mathematics. Kant, as we have seen, does not want to give this kind of necessity to final-cause thinking. This surely shows that Kant, and probably Cuvier also, was caught trying to define a notion of necessity for teleological thought that does not sit altogether comfortably with our thinking on necessity in other instances. Speaking now both as a historian and as a Darwinian, this does not surprise me. Sophisticated though it may be, the Kant/Cuvier position cannot ultimately be correct. Conversely however, this does not mean that it is philosophically worthless or scientifically without influence.

CONCLUSION

Cuvier was wrong about evolution. It does happen. But, following Kant, he was right to stress the design-like nature of the world—even to the point where it led him to deny the possibility of organic change. Following Kant he was right to stress that design is not some contingent thing, that nature might show irrelevant, but something deeply part of our understanding of the organic world and a real barrier to theories of change. Foucault was right to seize on this point, and to argue that it is this that sets the agenda for the evolutionists—first Charles Darwin (1859) and then his successors down to the present day. It was Darwin's genius to see that Cuvier (together with others—for instance, the British natural theologians like Archdeacon William Paley)[29] had set the questions, posed the problems that must be answered, as it was also Darwin's genius to find the solution, natural selection. Final cause is a barrier to theories of change, and Darwin broke down the barrier. And this takes us back to Immanuel Kant. Like Moses, he was never to enter the Promised Land—Israel for the one, evolution for the other—but he did lead us to the borders.

NOTES

1. C. Wilson, "Kant and the Speculative Sciences of Origins," in *The Problem of Animal Generation in the 17th and 18th Centuries*, ed. J Smith, (Cambridge: Cambridge University Press, 2006), pp. 375–401.

2. R. J. Richards, *The Romantic Conception of Life: Science and Philosophy in the Age of Goethe* (Chicago: University of Chicago Press, 2003).

3. A. O. Lovejoy, "Kant and Evolution" [1911], in *Forerunners of Darwin*, ed. B. Glass, O. Temkin, and W. L. Strauss Jr. (Baltimore: Johns Hopkins University Press, 1959), pp. 173–206; J. F. Cornell, "Newton of the Grassblade? Darwin and the Problem of Organic Teleology," *Isis* 77 (1986): 405–21.

4. M. Foucault, *The Order of Things: An Archaeology of the Human Sciences* (New York: Pantheon, 1970), p. 274.

5. I. Kant, *The Critique of Teleological Judgement*, trans. J. C. Meredith (Oxford: Oxford University Press, 1928 [1790]), p. 31.

6. Ibid., p. 25.

7. Ibid.

8. Ibid., pp. 65–66.

9. Ibid., p. 53.

10. Ibid.

11. Ibid., p. 54.

12. Ibid., pp. 78–79.

13. Ibid., 79n–80n.

14. Ibid., 80n.

15. I. Kant, *Opus Postumum*, ed. E. Förster (Cambridge: Cambridge University Press, 1993), p. 57.

16. Ibid., pp. 66–67.

17. Ibid., 85n–86n.

18. Richards, *Romantic Conception of Life*.

19. Kant, *Opus Postumum*, p. 122.

20. G. W. F. Hegel, *Philosophy of Nature* (Oxford: Oxford University Press, 1970 [1817]), p. 21.

21. L. Wright, "Functions," *Philosophical Review* 82 (1973): 160.

22. Kant, *Critique of Teleological Judgement*, p. 18.

23. G. Cuvier, *Le règne animal distribué d'aprés son organisation, pour servir de base à l'histoire naturelle des animaux et d'introduction à l'anatomie comparée* (Paris, 1817), pp. 1, 6, quoted in W. Coleman, *Georges Cuvier, Zoologist: A Study in the History of Evolution Theory* (Cambridge, MA: Harvard University Press, 1964), p. 42.

24. Ibid., pp. 67–68.

25. Ibid., p. 77.

26. Ibid., p. 92.

27. G. Cuvier, *Theory of the Earth*, 4th ed., ed. Robert Jameson (Edinburgh: William Blackwood, 1822, [1813]), p. 123.

28. Ibid., pp. 125–26, 29. W. Paley, *Natural Theology* (Collected Works: IV) (London: Rivington, 1819 [1802]).

Three

DARWINISM AND MECHANISM

mechanism: 1) the structure or adaptation of parts of a machine. 2) the mode
of operation of a process.

—*Concise Oxford Dictionary*

The discovery of the mechanisms by which causes produce or generate
their effects is a central part of a scientific investigation. The discoveries
of the mechanism of chemical reactions, of the mechanism of inheritance, and so many more, are examples of the fulfillment of this search.
But a word of caution is needed here as to the meaning of *mechanism*.
In ordinary English this word has two distinct meanings. Sometimes it
means "mechanical contrivance"—a device that works with rigid connections, like levers, the intermeshing teeth of gears, axles, and strings.
Sometimes it means something much more general, namely, any kind
of connection through which causes are effective. It is this latter sense
that the word is used in science generally, in such diverse expressions as
"the mechanism of the distribution of seeds" and "the mechanism of
star formation." In hardly any of these cases is any mechanical contrivance being referred to. So we must firmly grasp the idea that not all
mechanisms are mechanical.[1]

This is an essay about Charles Darwin, but my intent is as much
philosophical as it is purely historical. I am interested in the nature of science. Is science a disinterested reflection of objective reality or is it a
social construction, a subjective epiphenomenon on the culture of the

day? In previous publications, in search for an answer to my question, I have focused on the nature and role of metaphor in science.[2] Basically, I have argued that metaphor is widely used and indispensable. Moreover, it points to a middle way between the objective and the subjective. Science is objective, inasmuch as it is structured and guided by epistemic factors or values. It is beyond the individual or purely cultural because it aims to be predictive, consistent with other knowledge claims, internally coherent, unificatory, and simple. Yet science is in some way subjective, because we also structure and interpret it through our metaphors, things drawn from individual experience and the culture(s) within which science is produced. Nor are the metaphors readily eliminable, for—showing that in a sense the objective/subjective dichotomy collapses—they give rise to some of the crucial epistemic virtues of the best kind of science, the kind of science we most readily think of as objective. Most particularly, metaphor generates predictive fertility—or, as it is sometimes said, metaphor is the key factor behind the heuristic power of good theories. In this essay, I want to look at one of the most important and powerful metaphors in the history of science, the metaphor of "nature is a machine" (the mechanism metaphor) to see what role (if any) it played in the thinking of Charles Darwin and subsequent evolutionists, and to find what if anything this discussion contributes to my philosophical concerns.

NATURE IS A MACHINE

"Nature is a machine" is a metaphor that was introduced into science in the seventeenth century by people like the French philosopher-scientist René Descartes.[3] This metaphor was rooted in the culture of the day, namely that we can think of the world as a creation of God, and what God created was an efficiently functioning machine rather than (say) a plant or alternatively a totally useless and random mess. In other words, the metaphor was bound up with natural religion (the religion of reason), rather than revealed religion (the religion of faith). Metaphorically, one is thinking of the world as a machine; literally, one is thinking of the world as a creation by God. As intimated by the quotations given at the head of this essay, the metaphor fragmented into two senses: the world as a whole as a machine and individual parts as mechanical contrivances. As is often the case with fertile mechanisms, this led to something of a tension generated by the metaphor itself—machines are entities that work according to unbroken law and at the same time they are entities with an end, a purpose.

At the general level (leading to definition 2 in the dictionary, to Harré's "any kind of connection"), one starts to see the world as governed by unbroken law, regarding it as something like a clock, forever going in motions without end. The prime example of course is Newton's mechanics—provocative and pregnant term! This attitude (the "mechanical philosophy") is marked by (in Aristotelian terms) an exclusive reliance on efficient (and perhaps formal and material) cause, and a rejection of final causes. The English physicist Robert Boyle is the definitive source here. Against the Aristotelian notion that somehow "nature" itself has a being and a kind of mind or life force of its own, and making reference to a wonderful clock built (between 1571 and 1574) by the Swiss mathematician Cunradus Dasypodius, Boyle responded:

And those things which the school philosophers ascribe to the agency of nature interposing according to emergencies, I ascribe to the wisdom of God in the first fabric of the universe; which he so admirably contrived that, if he but continue his ordinary and general concourse, there will be no necessity of extraordinary interpositions, which may reduce him to seem as if it were to play after-games—all those exigencies, upon whose account philosophers and physicians seem to have devised what they call nature, being foreseen and provided for in the first fabric of the world; so that mere matter, so ordered, shall in such and such conjunctures of circumstances, do all that philosophers ascribe on such occasions to their omniscient nature, without any knowledge of what it does, or acting otherwise than according to the catholic laws of motion. And methinks the difference between their opinion of God's agency in the world, and that which I would propose, may be somewhat adumbrated by saying that they seem to imagine the world to be after the nature of a puppet, whose contrivance indeed may be very artificial, but yet is such that almost every particular motion the artificer is fain (by drawing sometimes one wire or string, sometimes another) to guide, and oftentimes overrule, the actions of the engine, whereas, according to us, it is like a rare clock, such as may be that at Strasbourg, where all things are so skillfully contrived that the engine being once set a-moving, all things proceed according to the artificer's first design, and the motions of the little statues that as such hours perform these or those motions do not require (like those of puppets) the peculiar interposing of the artificer or any intelligent agent employed by him, but perform their functions on particular occasions by virtue of the general and primitive contrivance of the whole engine.[4]

As it happens, Boyle himself was God obsessed. But was not his vision halfway to deism, to a being who set all in motion and who then stands back? Certainly, if you look at things with the advantage of hindsight, there is something to this. If you take soul out of the universe and substitute an unfeeling machine, you are indeed on the way to a godless creation—at least, to a creation that can be regarded as godless. Boyle himself was as much against deism as he was against Aristotelianism. He thought that God creates the universe and then holds it immanently in His hands. If He quits at any moment, then everything collapses. For this reason, Boyle was (unlike the deist) ready to accept miracles. God can (or did) do these just as He pleases, because He is involved all of the time. But, of course, the trouble begins when you start to find that what you thought was miraculous is in fact something that obeys the rule of law, a possibility that Boyle fully recognized, even though he thought it a point in his favor and against the Aristotelians:

> And when I consider how many things that seem anomalies to us do frequently enough happen in the world, I think it is more consonant to the respect we owe to divine providence to conceive that, as God is a most free as well as a most wise agent, and may in many things have ends unknown to us, he very well foresaw and thought it fit that such seeming anomalies should come to pass, since he made them (as is evident in the eclipses of the sun and moon) the genuine consequences of the order he was pleased to settle in the world, by whose laws the grand agents in the universe were empowered and determined to act according to the respective natures he had given them; and the course of things was allowed to run on, though that would infer the happening of seeming anomalies and things really repugnant to the good or welfare of divers particular portions of the universe.[5]

Yes, all of this is no doubt true, but the fact is that, as a need of a God of miracles recedes, then the need of a God at all recedes. The world is a clockwork and leave it at that.

For Boyle, it was at this point that the other (more specific) sense of the metaphor—parts of the world as mechanisms or contrivances (as Harré calls them, dictionary definition 1)—started to kick in. One focuses here on the purpose or point of a machine. This really is a teleological notion, one that puts final cause up front. Unlike Francis Bacon and Descartes who thought that they could do science without final causes—vestal virgins, Bacon called them, pretty but sterile—Boyle realized that in the biological world they are necessary for full understanding. If the eye is not made for seeing, then absolutely nothing makes sense at all. Against the Frenchman and his followers, there is a positive moral obligation to study nature and to work out its adaptations. As Boyle wrote in another of his essays, the "Disquisition about the Final Causes of Natural Things:"

> For there are some things in nature so curiously contrived, and so exquisitely fitted for certain operations and uses, that it seems little less than blindness in him, that acknowledges, with the Cartesians, a most wise Author of things, not to conclude that, though they may have been designed for other (and perhaps higher) uses, yet they were designed for this use. As he, that sees the admirable fabric of the coats, humours, and muscles of the eyes, and how excellently all the parts are adapted to the making up of an organ of vision, can scarce forbear to believe, that the Author of nature intended it should serve the animal to which it belongs, to see with.[6]

Boyle continued that supposing that "a man's eyes were made by chance, argues, that they need have no relation to a designing agent; and the use, that a man makes of them, may be either casual too, or at least may be an effect of his knowledge, not of nature's." But not only does this then take us away from the urge to dissect and to understand—how the eye "is as exquisitely fitted to be an organ of sight, as the best artificer in the world could have framed a little engine, purposely and mainly designed for the use of seeing"—but it takes us away from the designing intelligence behind it.

Boyle did not see this position of his as something threatening to the mechanical position but as complementing it. He thought that the general sense of *mechanism* (world as machine, process) was complemented by and harmonized with the specific sense of *mechanism* (parts as contrivances). But one can see why people like Bacon and Descartes—who theologically accepted final causes—wanted final causes out of science and were pushing toward a position of what in today's terms we might call methodological naturalism or mechanism, where the original concept of nature as a machine (the general sense) has become (what literary theorists call) a dead metaphor. As God gets further and further removed from science judged as science—and this has been the tendency since the seventeenth century—the very success of the metaphor of "nature is a machine" has brought about its demise. As scientists, we no longer think of nature as created by an intelligence. And this being so, as Harré rather hints, this means that although we might continue to talk in terms of mechanism, at this level, we do not really mean it. We do mean something and we do mean something very important—that nature works according to unbroken law—but we do not mean that nature is machine-like in the sense of "designed by an intelligence."

At this point, you might be led to conclude that, with the general sense of the metaphor now gutted of its original force, for all that Boyle argued otherwise, the practice of science puts pressure on the scientists to downplay or eliminate the specific sense. If nature is a machine only in name, then it is anomalous—if not outrightly contradictory—to speak of parts as machine-like, as contrivances or specific mechanisms. By stressing the clockwork nature of the world in its own right without reference to purpose, you are undermining the intent of using the clock to tell the time. The act of referring to intent is, to use an evolutionary analogy, rather like having an appendix: perhaps useful once but no longer. Even more, like having some feature that was useful but now is positively harmful, such as (in Western countries) a fondness for sweet things. And if this is all so, then looking at things from a broader philosophical perspective, we surely have support for the position of the objectivists about science. They might agree that in its earlier, immature stages, science is often if not always something much connected (primarily if not exclusively through metaphor) to the ways and norms of the society within which it is conceived. But they will crow that, as the science matures, the epistemic norms take over and expel the cultural and societal.[7] And a mark of this is that the metaphors become less significant and active—either they become moribund, as is the case of the more general "nature as machine," or they are expelled outright, as must or should have happened to "organic parts are contrivances." To quote the philosopher Jerry Fodor, "When you actually start to do the science, the metaphors drop out and the statistics take over."[8] Although science starts in culture, when it matures it really is objective and culture free.

DARWIN ON MECHANISM

This somewhat lengthy prolegomenon now sets us up to think about Charles Darwin. In his masterwork, *On the Origin of Species*, Darwin established the fact of evolution and proposed a causal force, that which today is taken as the key factor in change: natural selection. Surely, one might think, this is all going to fit very nicely and smoothly into the story just sketched above. Darwin strikes a key blow for the general "nature as a machine" metaphor, because he is showing how organisms come into being as a result of the workings of blind, unguided law. His is the methodologically naturalistic theory par excellence, and in this sense natural selection is the apotheosis of what the modern scientist means by mechanism—"connection through which causes are effective." But because he is so successful, Darwin is a major player in demoting or transforming this way of thinking into the status of dead metaphor. After Darwin, the world of organisms is no longer itself an organism (as the German idealists, the *Naturphilosophen*, would have argued) or a direct Creation (as natural theologians from Plato to William Paley would have argued) or a domain of vital forces (as Aristotle would have argued). It is just something that works by law, without final causes. And this being so, the other sense of mechanism, the specific sense, parts as contrivances, is doomed to oblivion. There are human-man mechanisms, of course—in Boyle's day, clocks; in our day, computers. But not in nature, and so we expect to find this kind of talk diminishing and ultimately vanishing.

In fact, things did not happen this way at all. With the help of concordances, printed and computerized, I have surveyed all of Darwin's major and several not-so-major works. I have looked also at all of the letters, published and unpublished. He simply does not speak of natural selection as a mechanism. He does not use the language of the "nature is a machine" general metaphor at all. This does not occur in the *Notebooks*, in the earlier essays (the *Sketch* and the *Essay*), in the *Origin of Species* (early or late editions), the *Descent of Man*, or the *Expression of the Emo-*

tions. He does not do so in the *Variation under Domestication*, a significant omission because, in his initial discussion in this work of the topic of selection, Darwin is at a point where he might naturally have done so. He is defending the use of the metaphor of natural selection, and he admits that his language is all rather anthropomorphic, inasmuch as it implies that the force of evolution is a power or an intelligent being. But, Darwin does not say that he is using selection as a mechanism or as something akin to a machine. He says simply, "I mean by nature only the aggregate action and product of many natural laws,—and by laws only the ascertained sequence of events."[9]

Given how much Darwin wrote—the many editions of the books, the papers, the thousands of letters that are only now being published—it would be a foolhardy person who claimed that Darwin never referred to selection (or any rival cause) as a mechanism, but one can say that this was absolutely not the typical language that he used about his theory in the prime creative years. (Amusingly, the editors of the online edition of the letters often refer to natural selection as a mechanism in their notes.) Moreover—and this is what is striking—Darwin's failure to speak of selection as a mechanism did not stem from his unwillingness to speak of mechanisms. From the beginning, from before he was an evolutionist, he was prepared openly to use the metaphor of mechanism in the specific sense, of mechanism as a contrivance. At times, his use is unselfconscious. At times, particularly when selection is on the table, his use is deliberate. Either way, he outdid Boyle. Thus in the *Voyage of the Beagle*, in a nonevolutionary context, we have:

> When we were at Bahia, an elater or beetle (Pyrophorus luminosus, Illig.) seemed the most common luminous insect. The light in this case was also rendered more brilliant by irritation. I amused myself one day by observing the springing powers of this insect, which have not, as it appears to me, been properly described. . . . The elater, when placed on its back and preparing to spring, moved its head and thorax backwards, so that the pectoral spine was drawn out, and rested on the edge of its sheath. The same backward movement being continued, the spine, by the full action of the muscles, was bent like a spring; and the insect at this moment rested on the extremity of its head and wing-cases. The effort being suddenly relaxed, the head and thorax flew up, and in consequence, the base of the wing-cases struck the supporting surface with

such force, that the insect by the reaction was jerked upwards to the height of one or two inches. The projecting points of the thorax, and the sheath of the spine, served to steady the whole body during the spring. In the descriptions which I have read, sufficient stress does not appear to have been laid on the elasticity of the spine: so sudden a spring could not be the result of simple muscular contraction, without the aid of some *mechanical* contrivance. [Here, and in subsequent quotations, I have italicized the language of machine or mechanism.][10]

Likewise in the *Barnacle* books:

Alcippe, according to Mr. Hancock, attacks only dead shells of the Fusus and Buccinum, and always on their inner sides, especially on the columella. The excavations, in the specimen which I examined, were so numerous as almost to touch, and sometimes to run into each other, the included animal being thus rendered distorted. The orifices are directed with respect to the shell indifferently upwards or downwards. From the shape and size of the cavity corresponding to that of the included animal, there can be no doubt, as stated by Mr. Hancock, that Alcippe forms its own cavity. That the action is *mechanical* I think may safely be inferred from the whole outer membrane being studded with minute, star-headed points of hard chitine, which rise from halo-like little discs of thickened membrane, which latter are well adapted to allow the underlying adherent muscular layer to act on the points, and thus on the surrounding shell.[11]

Notice that, as with Boyle, it is in the final cause (a term Darwin often uses) context, the adaptation context, that Darwin uses the language of mechanism. (*Mechanism* in the specific sense, that is definition 1 in the dictionary.) Since natural selection is intended to explain adaptation, we expect to find that it is in such contexts that Darwin will really use the metaphor—and he does. In the *Origin*, for instance, he sets up the problem in this language:

It may be doubted whether sudden and considerable deviations of structure such as we occasionally see in our domestic productions, more especially with plants, are ever permanently propagated in a state of nature. Almost every part of every organic being is so beautifully related to its complex conditions of life that it seems as improbable that any part should have been suddenly produced perfect, as that a complex *machine* should have been invented by man in a perfect state.[12]

And in the little book on orchids—the most important of all of Darwin's writings for showing how he thinks that selection will actually work—constantly he uses the machine metaphor to make his point:

> When I first examined these flowers I was much perplexed: trying in the same way as I should have done with a true *Orchis*; I slightly pushed the protuberant rostellum downwards, and it was very easily ruptured; some of the viscid matter was withdrawn, but the pollinia remained in their cells. Reflecting on the structure of the flower, it occurred to me that an insect in entering to suck the nectar, from depressing the distal portion of the labellum, would not touch the rostellum; but that, when within the flower, from the springing up of the distal half of the labellum, it would be almost compelled to back out parallel to the stigma by the higher part of the flower. I then brushed the rostellum lightly upwards and backwards with the end of a feather and other such objects; and it was pretty to see how easily the membranous cap of the rostellum came off, and how well, from its great elasticity, it fitted the object, whatever its shape might be, and how firmly it clung to it from the viscidity of its under surface. Together with the cap large masses of pollen, adhering by the threads, were necessarily withdrawn.
>
> Nevertheless the pollen-masses were not nearly so cleanly removed as those which had been naturally removed by insects. I tried dozens of flowers, always with the same imperfect results. It then occurred to me that an insect in backing out of the flower would naturally push with some part of its body against the blunt and projecting upper end of the anther which overhangs the stigmatic surface. Accordingly I so held the brush that, whilst brushing upwards against the rostellum, I pushed against the blunt solid end of the anther . . . ; this at once eased the pollinia, and they were withdrawn in an entire state. At last I understood the *mechanism* of the flower.[13]

And again:

> Still more interesting is this genus in its *mechanism* for fertilisation. We see a flower patiently waiting with its antennæ stretched forth in a well-

adapted position, ready to give notice whenever an insect puts its head into the cavity of the labellum. The female Monachanthus, not having pollinia to eject, is destitute of antennæ. In the male and hermaphrodite forms, namely Catasetum tridentatum and Myanthus, the pollinia lie doubled up, like a spring, ready to be instantaneously shot forth when the antennæ are touched; the disc end is always projected foremost, and is coated with viscid matter which quickly sets hard and firmly affixes the hinged pedicel to the insect's body. The insect flies from flower to flower, till at last it visits a female or hermaphrodite plant: it then inserts one of the masses of pollen into the stigmatic cavity. When the insect flies away the elastic caudicle, made weak enough to yield to the viscidity of the stigmatic surface, breaks, and leaves behind the pollen-mass; then the pollen-tubes slowly protrude, penetrate the stigmatic canal, and the act of fertilisation is completed. Who would have been bold enough to have surmised that the propagation of a species should have depended on so complex, so apparently artificial, and yet so admirable an arrangement?[14]

What is fascinating is that Darwin uses the metaphor to explain aspects of nature that are machine-like in distinctive or peculiar ways. In particular, he makes the point that nature has to make do with what it has rather than with what it would like to have. Hence, often, contrivances come out as though they were designed by (what the English would recognize as coming from) Heath Robinson or (what the Americans would recognize as coming from) Rube Goldberg.

Although an organ may not have been originally formed for some special purpose, if it now serves for this end we are justified in saying that it is specially contrived for it. On the same principle, if a man were to make a *machine* for some special purpose, but were to use old wheels, springs, and pulleys, only slightly altered, the whole machine, with all its parts, might be said to be specially contrived for that purpose. Thus throughout nature almost every part of each living being has probably served, in a slightly modified condition, for diverse purposes, and has acted in the living *machinery* of many ancient and distinct specific forms.[15]

NATURE AS MACHINE AND NATURE AS CONTRIVANCE

Contrary to expectation, Darwin does not use the general metaphor—natural selection as a mechanism—but he does use the specific metaphor—organic parts as mechanisms. How are we to understand this? Although it is interesting, I do not want to read too much into the failure to use the mechanism language for natural selection. We know that Darwin was highly sensitive to the philosophies of science of his day, and generally those whom he read were themselves not into the language of mechanism for the world taken as a whole.[16] In particular, people like John F. W. Herschel and William Whewell (to take the two who had the greatest influence on Darwin) were practicing and sincere Christians, thinking that force was a reflection of God's will. Although they were certainly in the Boyle tradition of thinking of nature as a systematic network of law-governed events, they were not in the business of flaunting their naturalism or mechanistic approach to nature. They would use machine-metaphor language when talking about issues in physics. Herschel, for instance, talks of the "division of fluids, in mechanical language, into compressible and incompressible."[17] But at a more general level, they talk of laws and causes (particularly *verae causae*, the best kinds of causes). Whewell particularly, arguing explicitly that organisms come through nonnatural forces, could never have used "mechanism" generically for causes or processes, if this was to encompass organic origins. Indeed, believing that each area of science has its own fundamental principles ("Ideas"), he was not about to allow full mechanical language and understanding for chemistry: "In attempting to advance a theory of Causes in chemistry, our task is by no means to invent laws of *mechanical* force, and collections of forces, by which the effects may be produced. We know beforehand that no such attempt can succeed."[18]

There is no question that Darwin went further than his teachers and mentors—further than anyone—in regarding nature as a machine. The metaphor was there, even if not the language. This was what Darwin was about. Not just chemistry, but biology also. To his friends Charles Lyell and Asa Gray, who were wriggling to find some kind of special forces for organisms—special forces that would guide evolution, and take the sting out of the law-like nature of history—Darwin was blunt to the point of rudeness.[19] There was to be no compromise here. It is law all the way down. Even chance is taken merely as a mark of ignorance rather than something ontologically or epistemologically significant. And as Darwin's own religious beliefs moved from the theistic Christianity of his youth, to a kind of deism (held when he became an evolutionist and right through the writing of the *Origin*), and on to agnosticism in his old age, I suspect he would have been happy to think of the metaphor as increasingly moribund and finally dead. The world is like a machine—definition 2 in the dictionary—but it isn't really one. Leave it at that.

More interesting is Darwin's use of the other sense of the metaphor (definition 1), the specific one that sees organic parts as machine-like, as mechanisms. This was language that he found in the writings of his mentors. Paley's *Natural Theology* was the biggest influence of all on Darwin in this respect, and opening randomly one finds, "Movable joints, I think, compose the curiosity of bones; but their union, even where no motion is intended or wanted, carries marks of mechanism and of mechanical wisdom."[20] Darwin takes this kind of language and thinking on board, fully. Of course, he transforms it with his explanation of natural selection, and this leads to new insights, for instance about the ramshackle nature of so many adaptations, doing with what they have rather than what they would like to have. But the very last thing Darwin wants to do is drop or belittle the metaphor.[21] Seeing nature's parts as machines, as mechanisms, as contrivances, is absolutely crucial for Darwin. Like a vampire before a virgin, the metaphor takes on new life. It is the key heuristic tool for the student of natural selection. One has the force. Now, how is one to apply it? Think of organisms and their parts as if they were machines, and puzzle out the solution. This is what Darwin does in his study of orchids and this is what he invites his readers to do. (Today, this is known revealingly as "reverse engineering.")

So, reverting again to the philosophical world, what would I want to conclude here? To make a full case for the thesis that the theory of evolution through natural selection is objectively true, one would need to run through the theory showing how it manifests epistemic excellence—that it is predictive, consistent, coherent, and so forth. Although I have tackled this issue elsewhere, here the claim will have to be taken on trust.[22] What can be said here is that the epistemic success of the general "nature is a machine" metaphor—leading to definition 2—and its increasing divorce from the notion of an actual creator underlines the truth of the thesis. Inasmuch as Darwin's theory participates in this metaphor—and it does—it is on the track toward objectivity. In fact, I would say that the success of evolution and natural selection as instruments of science (with respect to epistemic virtues like unification and predictive fertility) is a major reason why it is proper to conclude that science is really about something and not just a construction. (It is worth noting that Darwin himself was sensitive to the need for epistemic excellence and went out of his way to argue that his theory lived up to the demands. For example: "The present action of natural selection may seem more or less probable; but I believe in the truth of the theory, because it collects under one point of view, and gives a rational explanation of, many apparently independent classes of facts."[23]

At the same time, the ongoing importance of the metaphor of the machine in understanding contrivance (adaptation) suggests that the claim that science is purely objective—"knowledge without a knower,"[24] in Popper's felicitous phrase—is too strong a thesis. Live metaphors, rooted in culture, structure and inform our experience and are not about to vanish. In the case of Darwin's theory of evolution through natural selection, the vision of organic parts—bits of the orchid or of the barnacle—as machine-like, as mechanisms (definition 1), is absolutely vital to an understanding of how these bits or systems work. The very questions posed—how does fertilization in the orchid take place—are framed in terms of the metaphor of organic parts as machine parts. Why even ask about the bits and pieces unless you are trying to tie them together in a mechanical fashion? You see bits of metal on the ground, but unless you are thinking of them as parts of a watch or whatever, they make no sense at all. Likewise with the plants.

But notice that machines, particularly complex machines, are part of human culture—"Lucy" (*Australopithecus afarensis*) did not have them. Even if *A. afarensis* had been able to think conceptually (and Lucy surely could at some level), she could never even have asked Darwin's questions, let alone solved them. She did not live in a world of telescopes and watches and automobiles and computers. She could of course have asked questions about differential reproduction, and noted that some barnacles or orchids seem to survive and reproduce and others do not, but she could not have asked why the successful succeed and the unsuccessful fail. So she could have had an evolutionary theory, but by our standards it would have been impoverished. To get Darwin's theory, to get something so triumphant as objective science, you need the metaphors, and this brings in the subjective side of culture.

EVOLUTION AFTER DARWIN

But that was back then. What about now? Darwin was a pioneer. Perhaps, in the 150 years since the *Origin*, things have changed. Perhaps one side of the machine metaphor (the general side) was already back then on its way to its grave, and the other side (the specific side) would follow in the years subsequent? Actually however—and here I will move very quickly—this seems not to be the case. Certainly Darwin's use of mechanism persists. It was not long before people did actually start to use the language of mechanism about natural selection and other putative causes. This was part and parcel of the move to secularize science completely, and to use it as a tool of social reform and more. In the October 1864 installment of his *Principles of Biology*, Herbert Spencer wrote, "This survival of the fittest, which I have here sought to express in mechanical terms, is that which Mr. Darwin has called 'natural selection, or the preservation of favoured races in the struggle for life.'"[25] By 1873, Thomas Henry Huxley—who as the archetypical agnostic wanted nothing to do with God or with ideas of the world as His creation—was reducing everything to the blind workings of law, leading in a deterministic fashion to their inevitable ends: "The whole world, living and not living, is the result of the mutual interaction, according to definite laws, of the forces possessed by the molecules of which the primitive nebulosity of the universe was composed."[26] What we cannot now say is that the end to which all of this leads has any purpose. The universe works like a clock, but we have no right to say that the purpose is telling time any more than that the purpose is senseless ticking. In this world, the forces that control all of this—including all of this in the world of organisms—are mechanisms. "And there seems to be no reply to this inquiry, any more than to the further, not irrational, question, why trouble oneself about matters which are out of reach, when the working of the *mechanism* itself, which is of infinite practical importance, affords scope for all our energies?"[27] So it went. (It is for this reason that if someone discovered that Darwin in the 1870s referred to natural selection as a mechanism, I would take it as confirmation rather than refutation of what I have written above. By the 1870s, he was an agnostic, like—and under the influence of—Huxley, and if Darwin adopted the language of mechanism for causes, this would be part and parcel of the same general picture.)

As is well known, natural selection never gained much favor in Darwin's own lifetime. Around the beginning of the twentieth century there was a group, the biometricians, who became selection enthusiasts, but natural selection really had to wait until the 1930s before it started to take off as the generally acknowledged, central, causal force behind evolutionary change. By then, people—including religious people—referred unselfconsciously to selection and other putative causes as mechanisms.[28] Theodosius Dobzhansky, in his *Genetics and the Origin of Species*, tells his reader that "*mechanisms* that counter-act the mutation pressure are known to exist. Selection is one of them."[29] And again: "In its essence, the theory of natural selection is primarily an attempt to give an account of the probable *mechanism* of the origin of the adaptations of the organisms to their environment, and only secondarily an attempt to explain evolution at large."[30] (I suspect that this reading of the *Origin* would be news to its author.) More negatively from Sewall Wright, who was such an influence on Dobzhansky, "That evolution involves nonadaptive differentiation to a large extent at the subspecies and even the species level is indicated by the kinds of differences by which such groups are actually distinguished by systematists." Apparently only when you start to get up to the subfamily or family level do you start to get adaptive difference. "The principal evolutionary *mechanism* in the origin of species must thus be an essentially nonadaptive one."[31] More positively, over in England, we find Julian Huxley defending selection against various kinds of saltationism (evolution by macromutations) and pointing out that the "difference lies in the intermediary steps: in the one case the effect of use or function is supposed to be direct, in the other indirect, *via* the sifting *mechanism* of selection."[32]

Finally, jumping to the present, let us turn to the first page of today's standard text on evolution, *Evolutionary Analysis* by Scott Freeman and Jon C. Herron:

Where did Earth's organisms come from? Why are there so many different kinds? How did they come to be so apparently well-designed to live where they live and do what they do? These are the fundamental questions of evolutionary biology. The answers are found in both the pattern and the *mechanism* of evolution. The pattern of evolution is descent with modification from common ancestors. The principle *mechanism* that drives this change is natural selection.[33]

What about the more restricted use of mechanism as contrivance? Again, we can find that evolutionists in the Darwinian mode have used the metaphor right up to the present. Indeed, one of today's most influential thinkers, George Williams, has been explicit on this point:

> Whenever I believe that an effect is produced as the function of an adaptation perfected by natural selection to serve that function, I will use terms appropriate to human artifice and conscious design. The designation of something as the means or mechanism for a certain goal or function or purpose will imply that the machinery involved was fashioned by selection for the goal attributed to it. When I do not believe that such a relationship exists I will avoid such terms and use words appropriate to fortuitous relationships such as cause and effect.[34]

In this mode, when it comes to actual studies, the language of mechanism is used for adaptations. For instance, following standard Darwinian theory, David Reznick argues that fish that are heavily predated as adults will mature quickly, so that their reproduction can occur before they are killed and eaten. This is an adaptation to a specific situation. Fish that are eaten with less discrimination, and especially fish that are eaten young, will have no such selective pressure driving them this way. With great skill and subtlety Reznick was able to show that precisely these results hold for various little fish to be found in rivers on the island of Trinidad. Comparing two specific locations (high-predation patches were called generically "Chrenicichla" localities, and low-predation patches were called "Rivulus" localities), Reznick employed what is called the "mark-recapture" technique, where one simply catches all of the fish (in his case, guppies) in a particular location, marks them in some definite way so that they will not be confused with others, and then one releases them back.

Later one catches them all again—at least one catches again all of those remaining—and thus one can do comparative checks on the rates of predation. One can actually see if indeed the prey is being eaten by the predators.[35] "If the differences in predation caused differences in guppy mortality rates, then the recapture probabilities of guppies from Crenicichla localities should be lower. If the Crenicichla prey selectively on adults, then the difference in mortality should be more dramatic for the adult age classes. We did, indeed, find that the overall probability of recapture was substantially and significantly lower in Crenicichla localities, implying higher mortality rates."[36] As Reznick somewhat triumphantly concludes: "Such a result reveals a potential *mechanism* of life history evolution and thus goes a step farther in arguing that the differences in guppy life histories among Crenicichla and Rivulus localities represent an adaptation to predator induced mortality."[37]

I appreciate that not every evolutionist today is an ardent Darwinian, and those who are not are probably going to downplay the significance of selection and adaptation, and hence the importance of the "contrivance as mechanism" metaphor. My point is that those evolutionists who are Darwinians—and by far the biggest number of professional evolutionists do fall into this camp—follow Darwin himself. They accept and use the "contrivance as mechanism" metaphor, thinking it indispensable to their studies. Without it, they lose one of their most powerful heuristic tools.

CONCLUSION

My conclusion follows quickly and simply. Metaphor generally is important in science. The machine metaphor specifically has played a very important role in science—it is at the base of much of the epistemic triumph of science. Its more general reading—the world as a machine—is endorsed and accepted fully by the Darwinian evolutionist, although (as for other scientists) in important respects this is now a dead metaphor. The more specific reading—the parts of organisms are to be seen as mechanisms functioning for some end, created by and aiding natural selection—is absolutely crucial to modern evolutionary biology. This sense of the metaphor thrives mightily. It is an essential component of the predictive fertility of Darwinian evolutionary theory, and neither the past nor the present gives one reason to think that it will be eliminated. Thus, to repeat what I said at the end of the discussion of Charles Darwin's own work, the general reading of nature as a machine—nature as a lawbound system of events—lies beneath the justifiable claims of science to yield objective knowledge. This holds true of Darwin's own theory of evolution through natural selection, even more today than it did in his day. The specific reading of nature as a machine—organisms as composed of contrivances, as made of machine-like mechanisms—likewise thrives in today's (neo-Darwinian) evolutionary biology. In fact, indirectly it contributes to the epistemic excellence of the science. However, today no less than yesterday, the metaphor is one drawn from culture, and is not a necessary component of human thought or even of evolutionary thinking—although an evolutionary theory without the specific metaphor would be much impoverished. In this regard, because today's evolutionists do use the metaphor, neo-Darwinian science continues to have its subjective side. Given the success of the science, it would seem silly if the philosopher—in the name of some kind of epistemic purity—were to object to the way that things are. In the real world, science—the best science—reflects the human beings who create it.

NOTES

1. R. Harré, *The Philosophies of Science: An Introductory Survey* (Oxford: Oxford University Press, 1972), p. 118.

2. M. Ruse, *Monad to Man: The Concept of Progress in Evolutionary Biology* (Cambridge, MA: Harvard University Press 1996); M. Ruse, *Mystery of Mysteries: Is Evolution a Social Construction?* (Cambridge, MA: Harvard University Press, 1999); M. Ruse, *Darwin and Design: Does Evolution Have a Purpose?* (Cambridge, MA: Harvard University Press, 2003).

3. M. Ruse, "Robert Boyle and the Machine Metaphor," *Zygon* 37 (2002): 581–95.

4. R. Boyle, *A Free Enquiry into the Vulgarly Received Notion of Nature*, ed. E. B. Davis and M. Hunter (Cambridge: Cambridge University Press, 1996), pp. 12–13.

5. Ibid., p. 13.

6. R. Boyle, "A Disquisition about the Final Causes of Natural Things," in *The Works of Robert Boyle*, vol. 5, ed. T. Birch (Hildesheim: Georg Olms, 1966 [1688]), pp. 397–98.

7. E. McMullin, "Values in Science," *PSA 1982*, ed. P. D. Asquith and T. Nickles (East Lansing, MI: Philosophy of Science Association, 1983), pp. 3–28.

8. J. Fodor, "Peacocking," *London Review of Books*, no. 18, April 1996, p. 20.

9. I must acknowledge that there are those who argue that, far from adopting a mechanical view of the world, Darwin was deeply committed to an organicist picture of nature, being much indebted to the thinking of the German Naturphilosophen. This is the thesis of Robert J. Richards in his brilliant work *The Romantic Conception of Life: Science and Philosophy in the Age of Goethe*. I think Richards is just plain wrong on this point. Since Richards and I have just aired our differences in public, I will refer the reader to this clash. Basically, although I agree with Richards that Darwin was certainly aware of and responsive to German thinking, his real roots lay in his own country. I argue that the real influences on him were the industrialists of his day and their supporters. The wealth that supported Darwin and his wife, his first cousin, was founded on the use of machines and the socioeconomics (like that of Adam Smith) that enabled such machines to be used to the full, to fill the coffers of their inventors and owners. The machine-like view of the world, reinforced by his mother's and wife's Unitarian (and hence deistic) connections, was one that colored all of Charles Darwin's thinking about organisms and their origins. See R. J. Richards, "Michael Ruse's Design for Living," *Journal of the History of Biology* 37 (2004): 25–38, and M. Ruse, "The Romantic Conception of Robert J. Richards," *Journal of the History of Biology* 37 (2004): 3–23.

10. C. Darwin, *The Variation of Animals and Plants Under Domestication* (London: Murray, 1868), p. 6.

11. C. Darwin, *Journal of Researches into the Natural History and Geology of the Countries Visited during the Voyage of H.M.S. Beagle round the World*, 2nd ed. (London: John Murray, 1845).

12. C. Darwin, *A Monograph of the Fossil Balanidae and Verrucidae of Great Britain* (London: Palaeontographical Society, 1854).

13. C. Darwin, *On the Origin of Species by Means of Natural Selection, or the Preservation of Favoured Races in the Struggle for Life* (London: John Murray, 1859).

14. C. Darwin, *On the Various Contrivances by which British and Foreign Orchids Are Fertilized by Insects, and On the Good Effects of Intercrossing* (London: John Murray, 1862).

15. Ibid.

16. Ibid., p. 348.

17. M. Ruse, "Darwin's Debt to Philosophy: An Examination of the Influence of the Philosophical Ideas of John F. W. Herschel and William Whewell on the Development of Charles Darwin's Theory of Evolution," *Studies in History and Philosophy of Science* 6 (1975): 159–81.

18. J. F. W. Herschel, *Preliminary Discourse on the Study of Natural Philosophy* (London: Longman, Rees, Orme, Brown, Green, and Longman, 1830), p. 225.

19. W. Whewell, *The Philosophy of the Inductive Sciences*, vol. 2 (London: Parker, 1840), p. 264, his italics.

20. M. Ruse, *The Darwinian Revolution: Science Red in Tooth and Claw* (Chicago: University of Chicago Press, 1979).

21. W. Paley, *Natural Theology* (Collected Works: IV) (London: Rivington, 1819 [1802]), p. 86.

22. Ruse, *Darwin and Design*.

23. See M. Ruse, *Darwinism Defended: A Guide to the Evolution Controversies* (Reading, MA: Benjamin/Cummings, 1982); Ruse, *Mystery of Mysteries*; M. Ruse, *Darwinism and Its Discontents* (Cambridge: Cambridge University Press, 2006).

24. Darwin, *Variation of Animals and Plants*, pp. 13–14.

25. K. R. Popper, 1974, "Darwinism as a Metaphysical Research Programme," in *The Philosophy of Karl Popper*, ed. P. A. Schilpp, vol. 1 (LaSalle, IL: Open Court, 1972), pp. 133–43.

26. Herbert Spencer, *Principles of Biology*, vol. 1 (London: Williams and Norgate, 1864), pp. 444–45.

27. T. H. Huxley, *Critiques and Addresses* (New York: Appleton, 1873), p. 305.

28. Ibid., p. 307.

29. I do not know who first actually called natural selection a "mechanism." I would not be surprised to find that it was an American, although expectedly (given that he was an evangelical Presbyterian) Asa Gray never uses this term of selection in his collected essays, *Darwiniana*. (As expectedly, he constantly likens contrivance to machine-like processes.) In 1897, in *The Survival of the Unlike: A Collection of Evolution Essays Suggested by the Study of Domestic Plants*, 2nd ed. (New York: Macmillan), Liberty Hyde Bailey refers to natural selection as a hypothesis about the "controlling process or factor in evolution" (p. 57), although he does use the language of machinery to talk about heredity (p. 64). However, in his survey of evolutionary theories of 1907, *Darwinism Today* (New York: Henry Holt), Vernon L. Kellogg tells us that he is after the "factors and mechanism of organic evolution" (p. iii), and at the end of the book concludes that "Darwinism as the all-sufficient and even most important causo-mechanical

factor in species-forming and hence as the sufficient explanation of descent, is discredited and cast down" (p. 374). A quick search of the online works of the pragmatists C. S. Peirce and William James (including Peirce's *Pragmatism* [Indianapolis: Hackett, 1982] and James's *Varieties of Religious Experience: A Study in Human Nature* [New York: Longman, 1902]) does not suggest that the philosophers would have been key influences on the biologists. I find that the English at this time do not generally use the term mechanism to refer to selection. Certainly (covering myself against an atypical occurrence) one can say that the general trend is not to use the term in this sense. This is true of the Darwinian book *The Colours of Animals*, by the Oxford biologist E. B. Poulton (London: Kegan Paul, Trench, Truebner, 1890). Checking a little book written in 1912, *The Evolution of Living Organisms* (London: T. C. and E. C. Jack), that (to the best of my knowledge) first synthesizes selection and Mendelian genetics in a full fashion, I find that the author (the English anatomist E. S. Goodrich) refers to Mendelism as a "mechanism," but calls selection a "factor" or "process." Perhaps as people became more and more comfortable in seeing Darwinian selection and Mendelian genetics as complements, they decided that sauce for the Mendelian goose should also be sauce for the Darwinian gander. But even R. A. Fisher in his *Genetical Theory of Natural Selection* (Oxford: Oxford University Press, 1930) and J. B. S. Haldane in his *Causes of Evolution* (New York: Cornell University Press, 1932) do not come right out with the language.

30. T. Dobzhansky, *Genetics and the Origin of Species* (New York: Columbia University Press, 1937), pp. 37–38.

31. Ibid., p. 150.

32. S. Wright, "The Roles of Mutation, Inbreeding, Crossbreeding and Selection in Evolution," *Proceedings of the Sixth International Congress of Genetics* 1 (1932): 364.

33. J. S. Huxley, *Evolution: The Modern Synthesis* (London: Allen and Unwin, 1942), p. 39.

34. S. Freeman and J. C. Herron, *Evolutionary Analysis*, 3rd ed. (Englwood Cliffs, NJ: Prentice-Hall, 2004), p. 1.

35. G. C. Williams, *Adaptation and Natural Selection* (Princeton, NJ: Princeton University Press, 1966), p. 9.

36. D. N. Reznick, M. V. Butler IV, and H. Rodd, "Differential Mortality as a Mechanism for Natural Selection in the Guppy (*Poecilia reticulata*)," *Evolution* 50 (1996): 1651–60.

37. D. N. Reznick and J. Travis, "The Empirical Study of Adaptation in Natural Populations," in *Adaptation*, ed. M. R. Rose and G. V. Lauder (San Diego: Academic Press, 1996), p. 269.

Four

KICKING AGAINST THE PRICKS

Alfred Russel Wallace the Rebel

rebel: To resist or defy an authority or a generally accepted convention.

(American Heritage Dictionary)

Alfred Russel Wallace discovered natural selection, the driving force of evolution, in 1858. Yet within ten years he was arguing that selection could not possibly account for the evolution of humankind and that only spirit forces could have done the job. How could this possibly be? Had Wallace, the ultimate man of science, turned into Wallace, the ultimate rebel against science? This is my question. To answer, we need first some background.[1]

MID-VICTORIAN THINKING ON THE NATURE OF SCIENCE

The first part of the nineteenth century, in Britain the 1830s particularly, was an incredibly active and fertile period for the development of what we today would call the philosophy of science.[2] This was not a purely intel-

73

lectual phenomenon, especially not in England. Right through the first half of the century, with few exceptions, most of the science had to be done by people at Oxford and Cambridge, whose main income came from undergraduate teaching in nonscience subjects and who necessarily had to be ordained members of the Anglican Church. Hence, appreciative of the great importance of science and technology in an industrial society, there was a major effort even by (especially by) these people to find ways in which science could be supported in a secular setting. The founding of organizations like the British Association for the Advancement of Science, an annual gathering (always in a different city, outside London) of professionals and amateurs, was a significant part of this drive.[3]

Important here were the astronomer John F. W. Herschel;[4] William Whewell,[5] successively professor of mineralogy and professor of moral philosophy at the University of Cambridge (and inventor of the word "scientist"); and Baden Powell[6] (father of the scout master) at Oxford. They were articulating the role of the scientist; indeed the very meaning of what it is to be a scientist was being defined. Thanks to them and their friends, new scientific societies were being formed—notably the annual British Association for the Advancement of Science—and thinkers were turning their attentions to articulating the rules and methods and standards of science, most particularly of good science. Isaac Newton was the paradigm. His work, particularly his mechanics, was the ideal to which scientists should aspire. This meant above all that one should explain through unbroken law, particularly through deductively connected networks: axiom systems, or what we today would call hypothetico-deductive systems. One should also strive to be causal. One should aim for what Newton had called verae causae, true causes.

A major topic of discussion was the connection—or rather the nonconnection—of science and religion.[7] Socially and intellectually it was becoming important to separate these. Everyone agreed that science could have implications for religion. Anglicans had always stressed natural theology, and a major reason for promoting science was the way that it proved and glorified God. The 1830s was the decade of the *Bridgewater Treatises* (Whewell wrote one) devoted to the evidence of divine design in nature. However, everyone also agreed that it is inappropriate to bring God into one's science. By this time, by and large all with pretensions to being real scientists were no longer trying to prove such things as Noah's Flood, and even if they had been it would have had to be natural. The major problem was the origin of organisms. No one could see how their adaptive nature, their design-like nature, could come through unbroken law. Some like Herschel thought that there simply had to be as-yet-undiscovered natural laws. Whewell inclined more to miracles, but he was adamant that any claims of this nature would not be scientific. "The mystery of creation is not within the range of her legitimate territory; she says nothing, but she points upwards."[8]

One should say that the efforts of these people—even though as they themselves got older they got more conservative—were very successful. By the 1850s, you could start to make your way as a full-time scientist, without need or support of the church. It was not easy, but it could be done. Darwin's great supporter Thomas Henry Huxley got a university job in the mid-1850s, and always resisted going to Oxford or Cambridge, institutions that themselves were now introducing science degrees and starting to loosen the links with the church.

CHARLES DARWIN AND EVOLUTION

Against this background, still setting the scene for Wallace, turn now to the other discoverer of natural selection, Charles Darwin. Darwin (born in 1809, fourteen years before Wallace) was always interested in science. When at Cambridge he knew the chief scientists (notably the geologist Adam Sedgwick, the botanist John Henslow, and the already-introduced polymath William Whewell). He spent five years (1831–1836) on the HMS *Beagle* going around the world, not just collecting but working hard on the theories involved (especially geology). He worked nonstop on science when he returned, writing on geology, then on barnacles (an eight-year study), and many other topics, especially botanical (orchids, climbing plants, insectivorous plants, and more). Around 1850, Huxley—who liked to do these sorts of things—rated English biologists, and gave Darwin top marks. This was before he knew Darwin and almost a decade before the *Origin*.

Darwin became an evolutionist in the spring of 1837 and discovered natural selection in the fall of 1838, more than two decades before the *Origin* appeared in 1859. Darwin was a genius, and it was his genius that led to his great scientific achievements. But in context, his work is fully understandable and not so very surprising. Paradoxical as it might seem, it would be quite inappropriate to speak of Darwin as a "rebel." Quite apart from the fact that there was no way in which Darwin was going to alienate important scientists by revealing his thinking on evolution—he did not do this for twenty years until he was forced into doing so, by which time he had built around himself a group of supporters—his thinking flowed naturally from his background knowledge and training. Darwin made major moves forward, but not as a rebel. He just did not "resist or defy an authority or a generally accepted convention." For a start, Darwin knew that the origins of organisms were the big problem in biology. On the *Beagle* voyage, he read Lyell's *Principles of Geology*, a work that insisted that organic origins had to be natural and ongoing, and yet that dared not grasp the nettle of evolution (or anything else, for that matter). Then Darwin knew all about evolutionary ideas. His grandfather Erasmus Darwin was an evolutionist, and Charles had read his major work, *Zoonomia*. He had talked about evolutionary ideas with someone (Robert Grant) when he spent two pre-Cambridge years in Edinburgh, in an abortive attempt to become a doctor. In the *Principles*, Lyell talked about the French evolutionist Lamarck. And the leading scientist of the day, the astronomer John F. W. Herschel, had labeled the organic origins question as the "mystery of mysteries."[9]

Having moved to evolution, primarily because that was the only way he could sensibly explain the distribution of the reptiles and birds on the Galapagos Archipelago, Darwin knew that he had to find a cause and to be able to present it as a true cause. Darwin fell straight into artificial selection as a means of organic change, thanks to his rural connections (especially the breeding work of Josiah Wedgwood the younger, Darwin's uncle and about-to-be father-in-law). For Darwin, artificial selection, which we see and control, is analogous to natural selection—if the former is a cause, then it is reasonable to think that the latter is a cause. Then, after he had hit on the mechanism of natural selection, Darwin knew all about how to embed his ideas into a well-formed theory, using the mechanism to explain different areas like instinct, paleontology, biogeography, systematics, morphology, and embryology. Again, something bolstering the causal pretensions of natural selection. Something making one think that selection is a true cause. On top of this, of course, Darwin spoke to the design question. Because of his Cambridge training and nonstop diet of natural theology from the works of Archdeacon William Paley, Darwin knew that the defining feature of organisms is their adaptedness—the fact that they exhibit final cause. This is precisely what natural selection addresses.

It is important to pick up on this last point. When Darwin wrote the *Origin*, he was not an agnostic (this came in later years). Although no longer a Christian, he believed in a God who is an unmoved mover. He was a deist. And as such, he believed in God as a designer. To his American friend Asa Gray, just after the publication of the *Origin*, he wrote:

> I see no necessity in the belief that the eye was expressly designed. On the other hand I cannot anyhow be contented to view this wonderful universe & especially the nature of man, & to conclude that everything is the result of brute force. I am inclined to look at everything as resulting from designed laws, with the details, whether good or bad, left to the working out of what we may call chance.[10]

But not in science! When Gray wanted to introduce directed variations to supplement selection, Darwin simply said that Gray was no longer doing science. The pupil of the philosophers had learned the message well. There is no place for God in science.

VICTORIAN JONAH

Let us turn now to Alfred Russel Wallace. He was born to English parents just outside the town of Usk in Monmouthshire, Wales, on January 8, 1823. He left school early and, with some time out for school teaching, worked for a number of years as a surveyor—an occupation then much in demand given the huge growth and spread of the railways. In 1848, with his friend Henry Walter Bates, he left for the Amazon, where he was to spend some four years collecting specimens (chiefly insects) that were then sold to wealthy collectors back in England. After a couple of years in the old country, he set off again in 1854 for the East, and spent some eight years collecting on and around the Indonesian Archipelago. After he returned again to England, he settled down and married and raised a family. He never held a steady job, making a living from writing, from acting as an examiner on student tests, and eventually on a small government pension. He died on November 7, 1913, and was buried in Broadstone Cemetery in Dorset, England.

If one had to use one word to sum up the life and achievements of Wallace, *unfortunate* springs at once to mind. His troubles started before his birth, when his father (a solicitor) lost all of a comfortable inheritance in bad investments, and (in his son's words) achieved considerable relief in his situation from the fact that his affairs were so bad that they could sink no further. Then there is the missing *l* at the end of Wallace's middle name, a mistake made at the time of christening and destined to cause scholars infinite troubles for years to come. On his return from the Amazon, his ship burned down and his collections and papers were all lost. Things were little better when he returned from the East. Wallace thought he had found the perfect spouse until one day the young woman he was courting sent her father to the front door to tell the would-be suitor that he was never again welcome. He was "very hurt."

Wallace followed in his father's footsteps by losing his money from his collections, first by giving much to his brother-in-law's failing photography business, and then by investing in lead mines just before the silver mines in the American West discovered that their byproduct was huge amounts of cheap lead. He responded to a madman's challenge that the earth was flat and then, having successfully performed an experiment showing that the earth is round, discovered that in England wagers are not enforceable by law and had to spend huge amounts of money fending off the crazy flat-earther. Most unfortunate of all, of course, Wallace discovered the mechanism of evolution through natural selection, sent it off to one of his correspondents in England, the naturalist Charles Darwin, only to discover that Darwin had been sitting on that very idea for twenty years. Darwin went on in the next year (1859) to publish *On the Origin of Species by Means of Natural Selection*, and forever more the theory has been known as Darwinism rather than Wallaceanism. (Wallace himself wrote a book called *Darwinism*, published in 1889.)

WALLACE AND HIS ENTHUSIASMS

Let us dig a little more into Wallace's nature, and to do this, let us play him off against Darwin. The contrast could not be greater. Wallace was in the middle class but very barely. He went off collecting as a job, whereas Darwin's collecting was the hobby of a toff. Darwin did not have to sell his finds to keep himself in funds. Wallace was unfortunate, but often—all too often—his misfortune was of his own making. He may have been generous to relatives, but he was stupid to have thrown away the monies from his labors in the East. He was even more stupid to get involved in the flat-earth question. No sane man—no Darwin or Huxley—would have soiled his hands in this way. There was no wonder that no one ever wanted to give Wallace a job. Bates, when he returned, thanks to behind-the-scenes efforts by Darwin, was slotted into the secretaryship of the Geographical Society, and there he became indispensable. One would have had to be as silly as Wallace to have given him the post. It was not that people disliked Wallace. They were happy to pressure the prime minister (Gladstone) to get Wallace a pension. They just did not trust his judgment.

Above all, there were Wallace's enthusiasms. He could probably have been forgiven for his socialism and his endorsement of land nationalization—the latter a very popular movement in both Britain and America in the last decades of the nineteenth century, and supposedly a failsafe method of getting more equality into society. Darwin was a capitalist and Huxley admitted that by temperament he was a conservative, but there were others who inclined more to the left on these matters. What did stick in people's craws was that in the 1860s Wallace became an ardent spiritualist, believing that there are unseen forces that interfere in the plans and ways of humankind. As a result, Wallace wrote extensively on the subject, badgered people to go to séances, and appeared over and over again offering public testimonials to mediums who were accused of fraud.

The point I am making is that there was somewhat of the crank about Wallace. It will not surprise you to learn that he was antivaccination, or that he was tempted to vegetarianism—a temptation firmly squelched by his physician who put him on a diet of raw, finely chopped beef! My own favorite is Wallace's claim that the future of the human race lies in the hands—or rather the wombs—of young women. Apparently in the future, we will all rise upward because young women will choose as mates only the best and finest of young men, those worthy of love and respect:

> In such a reformed society the vicious man, the man of degraded taste or
> of feeble intellect, will have little chance of finding a wife, and his bad

qualities will die out with himself. The most perfect and beautiful in body and mind will, on the other hand, be most sought and therefore be most likely to marry early, the less highly endowed later, and the least gifted in any way the latest of all, and this will be the case with both sexes.[11]

One can only suppose that if the Wallace children truly behaved that way, they must have been as odd as their father. Not until the rise of American liberal arts colleges and their codes of conduct was the world again to see such a misunderstanding of human nature and of the sexual interests of the young.

WALLACE THE EVOLUTIONIST

If Wallace was so out of touch with reality, if he was so very different from Darwin, how then was he able to become an evolutionist and to hit on natural selection? Let me stress that we are not mistaken in his achievements. He did become an evolutionist and although there are some differences between the early thoughts of Wallace and Darwin— the former always had a predilection for group selection in a way that was alien to Darwin's thinking—the two men did hit on exactly the same mechanism. (I have discussed the Darwin-Wallace differences over the levels of selection elsewhere.)[12] They both saw that there is an ongoing struggle for existence, and that randomly appearing features that help in the struggle will tend to be preserved and that this will lead eventually to full-blown evolution. The similarity of the discoveries cannot be denied.

In a way, the important question is not so much the discovery of natural selection (in 1858) but the move to evolution (around 1845). I do not think that simply using the word *genius* is a satisfactory explanation of their respective discoveries of natural selection—that would be rather like Molière's joke explanation of a sleeping potion having a dormative virtue—but the fact is that for both Darwin and Wallace there was genius involved in finding natural selection. However, the genius was working in a pattern and with tools at hand. In finding natural selection, both Darwin and Wallace were by now committed evolutionists and searching for a cause. Both Darwin and Wallace read Robert Malthus's work on population and realized that organisms are in an ongoing struggle for existence. Both Darwin and Wallace realized that the features organisms have help in the struggle for existence. And so forth. In other words, there is nothing so very mysterious about the discovery of natural selection, although one can and should stand in awe of the abilities of Darwin and Wallace to put things together and get a solution.

I have shown already how we can start to make sense of Darwin's move to evolution if we take into account the way in which his group was stressing the organic origins problem, how he knew already about evolution as a putative explanation, and how he found things on the *Beagle* voyage that called out for an evolutionary history. I am not pretending that one could have predicted with certainty that Darwin would have become an evolutionist, but that he did is ultimately not so very mysterious. Wallace is different. He had nothing that Darwin had—no training in science, no group stressing the interest of the problems, and certainly in 1845 absolutely no firsthand experience of peculiar distributions of organisms in either space or time. Yet he became an ardent evolutionist. Why?

We can find an answer if, as is so often the case with these sorts of things, we take a holistic approach. We should not compartmentalize Wallace the scientist and Wallace the crank. Wallace was one of those people who by temperament are always attracted to outrageous positions, particularly if these positions are disliked by the respectable. The move to spiritualism was a classic example of the way that Wallace's mind worked. Paradoxically, Wallace rather liked it precisely because it was a laughed-at position. The same is true of land nationalization. He enjoyed kicking against the pricks. The conversion to evolution fit exactly into this pattern. In 1844, scandalizing the respectable scientists like Sedgwick and Whewell, an anonymously authored evolutionary tract appeared, *The Vestiges of the Natural History of Creation*. Written, as we now know, by the Scottish publisher Robert Chambers, it argued that all of life came from inorganic matter and then progressed steadily upward to humankind—to Queen Victoria, actually. It was an appalling work— weak on science, speculative beyond compare, verging on the heretical. In short, if Newton were the filet mignon, it was the fast food of science, and like all fast food was taken up with enthusiasm by the hoi polloi. Half of the fun was precisely that the respectable scientific community loathed it. Wallace read it with great enthusiasm and at once became a convert.

It was a minority position, it involved sweeping hypotheses, and it was sneered at by the knowledgeable.[13] There was just no way that Wallace could have resisted.

Being gullible and ignorant was precisely what went into Wallace's move to evolution. It was absolutely right and proper that no serious scientist accepted the evolutionary message of *Vestiges* on the evidence that it presented. Here, in case you are still worrying about how someone like Darwin could have become an evolutionist, I stress that the evidence of *Vestiges* was not the evidence of Darwin. The older evolutionist had fossils and geographic distributions—the tortoises and birds of the Galapagos—and artificial selection and top-quality embryology and more. *Vestiges* had none of this, or where it had something (like embryology) it had it wrong. It argued that frost ferns left on windows is evidence of spontaneous generation, that insects come out of electrical experiments, and that birds spontaneously change into mammals. It was not something for the serious scientist. But for a young man who loved outlandish ideas, especially if they were antiestablishment (and Wallace was just then getting very excited about socialism), *Vestiges* was gospel, to use a metaphor. Had Wallace not been childlike in his trust of the absurd, he would never have become an evolutionist.

What does make Wallace quite exceptional is that when he had an idea or a conviction, he was not about to give it up in the face of opposition from the respectable and he was prepared to labor for it. He may have been a crank, but he was a dedicated, hardworking crank. He spent years attending séances and supporting spiritualism and spiritualists. He spent over thirty years as president of the British land nationalization organization. Alfred Russel Wallace was no fair-weather friend. In the case of evolution, having become a convert he set about putting his convictions into action. Going to the Amazon was in major part fueled by his felt need to find real evidence for evolution. It was not simply a get-rich project. (Just as well, given how things turned out!) Then he kept thinking and thinking about the issues, trying to stay abreast of ideas back in Britain. In 1855 he published a seminal paper in which he argued that organisms always seem to appear in just the places on the globe where similar organisms already exist. If not overtly evolutionary, it was as close as it possibly could be. This led to correspondence with Darwin and explains why, of all people, it was Darwin to whom Wallace sent his selection paper. (Apparently they may have met earlier in the decade, but neither seems to have remembered any such encounter.) Wallace long knew of Darwin's travel book based on experiences from the *Beagle* voyage, and this was one of the inspirations sending him to the Amazon.

HUMANS

I have said that Wallace was a crank. Does this at once imply that he was a rebel? Not necessarily. Often cranks do become rebels, but one can imagine someone with many ideas that one would think of as crankish but not necessarily think of such a person as a rebel. If one grew up in a family with odd ideas about religion or diet or dress, one might think of such a person as a crank but agree that within the context that person was not particularly rebellious. This is a point that applies to Wallace. Had Darwin in the late 1830s come out in favor of evolution, although he might have had good reason to do so—he did have good reason to do so—socially he would have been a rebel. Wallace did not belong to this group—the group of professional scientists. He was part of the great unwashed. He was part of the half-informed, the ill-educated, the pandering-to-any-daft-idea brigade. For this reason, within his group—set in context—it is not appropriate to think of the Wallace who became an evolutionist in the mid-1840s as particularly rebellious. But now the question arises of the Wallace of the 1860s, the Wallace who denied that natural causes could explain human origins. Was this a case of rebellion? Let us pick up the story again.

I have said that Wallace was a crank. What I am not trying to insinuate is that Wallace was stupid. He hit on natural selection when people like Huxley did not. In the 1860s, he and Darwin had a very sophisticated discussion about the levels of selection. Although in respects I think Darwin was right and Wallace was wrong, the quality of the debate was not to be equaled again until the work of people like William Hamilton and George C. Williams in the 1960s.[14] And in the 1870s, Wallace did major and important and lasting work on biogeography. Wallace's line, dividing Asian and Australasian animals in the Malay Peninsula, is still important.[15] What I do want to say is that Wallace was an untutored crank. Or rather he was a self-taught crank. He had not had the university education of Darwin, nor had he been directed to the books of the philosophers—he had not come to intellectual maturity among a group of men who were trying desperately to articulate the nature and rules of good science. Indeed his education generally was very spotty. When Wallace tried his hand at school teaching, each night (preparing for the next day's class) the principal had to help him understand the material. This lack of training, this lack of scientific sophistication showed, and nowhere more so than in the case of humans in the 1860s. The story is simple. At first, Wallace was eager to offer a natural, selection-driven account of human origins.[16] He did this, pointing out how intelligence was going to be something really important in the case of human evolution, and that it would be very significant in the struggle for existence. It would be a major adaptation. Darwin for one was very excited about this thinking, writing to praise Wallace strongly.

Then, as Wallace got enamored with spiritualism, he reversed himself. A natural explanation of human origins was no longer enough. One had to accept that spirit forces were directing our history. In support of this, Wallace instanced a number of features—hairlessness for one, our great intelligence for another—that simply could not have come through selection.[17] Savages have the potential for great thoughts but rarely if ever have them. So the brain cannot have exploded upward in size because it was an adaptation. Another thing—or rather, another Thing—must have been responsible. Darwin of course was appalled and accused Wallace of killing their child. Darwin was also now spurred to write on humankind himself, and this led in 1871 to the publication of the *Descent of Man*. Offering a completely naturalistic account of human origins, the one concession Darwin made to Wallace was to agree that things like intelligence could not be produced by simple natural selection. For this, Darwin relied on his secondary mechanism of sexual selection, which involves competition within species for mates. Clever people apparently have more kids. Powerful men get the trophy wife.

Why on earth did Wallace start appealing to spirit forces? Break this down into two questions. First, why did Wallace accept spiritualism? Second, why did he think he could use it to explain human origins, or rather, why did he think (as he certainly did) that other scientists should accept spiritualism to explain human origins? The answer to the first question is easy. Wallace accepted spiritualism for the usual reasons—he thought that the table lifting, and the knocks and shrieks, and the rest of the rather weird (and to modern-day thinkers, not to mention many of his contemporaries, rather pathetic) phenomena could be explained only by invoking the supernatural. For all the many cases of fraud, Wallace thought there was enough to justify belief. He simply refused to accept the criticisms of friends and of those not so friendly—for instance about being unable to repeat things in a controlled fashion, or allowing observers to sit where they wanted rather than where the medium wanted. In a truly ecumenical fashion, Wallace accepted the miracles at Lourdes, and almost took comfort from the fact that such miracles seem to be random and have nothing to do with merit or gravity of disease or whatever.

More than this, Wallace refused to balance spiritualism against the great body of evidence that the world simply does not work that way. Sounds on an everyday basis are produced by physical things clashing together and not by forces from the other world. Tables levitating have pulling strings or pushing fingers, not the dear departed moving into the furniture business. John Stuart Mill said flatly to Wallace, "For my own part I have not only never seen any evidence that I think of the slightest weight in favour of spiritualism, but I should also find it very difficult to believe any of it on any evidence whatsoever, and I am in the habit of expressing myself to that effect very freely whenever the subject is mentioned in my presence."[18] Wallace's response to this was that Mill was "very unphilosophical." But of course in the eyes of Mill and of the others, including Darwin, it was Wallace who was being unphilosophical. It was Wallace who was refusing to abide by the rules of science—controlled experiment and so forth, especially for something that is going against the huge warehouse of experience. It was not (to put things in a contemporary idiom) that Wallace and only Wallace was a true Popperian, prepared to put the most favored hypothesis (that general opinion is wrong) to the test, but that Wallace himself was no true Popperian for he would not put his thinking (in favor of spiritualism) to the test.

Why did Wallace behave like this? Obviously partly for the kinds of psychological reasons already discussed. But as significant, if not more, was the fact that Wallace had never truly internalized the rules of science. He was a man outside the system, and while he had brilliant flashes of intuition, in respects his thinking was never disciplined. He did not have

Darwin's training nor (to think of people from the next cohort) did he have the social position and responsibilities that Huxley had in the world of science. It is fascinating and surely pertinent to learn that another evolutionist who embraced spiritualism with fervor was Robert Chambers, the still-anonymous author of the *Vestiges*! There were happy exchanges of letters. "I have for many years known that these phenomena are real, as distinguished from impostures; and it is not of yesterday that I concluded that they were calculated to explain much that was doubtful in the past, and when fully accepted, revolutionize the whole frame of human opinion on many important matters."[19] Chambers like Wallace was in respects not a professional scientist—on the fringes of the business—and so had not internalized the rules.

The conclusion to Chambers's letter to Wallace (written just about the time that Wallace was going to come out on humans) gives a clue to answering the second question posed above, the question about why Wallace plunged ahead and claimed that his position on humans was scientific: "My idea is that the term 'supernatural' is a gross mistake. We have only to enlarge our conceptions of the natural, and all will be alright." This is the key point. Chambers—and obviously Wallace—were simply not prepared to separate off the natural from the supernatural—the properly scientific from the properly religious. It was not that Wallace wanted to say, as Whewell might have said in the 1830s, that we simply cannot give a scientific explanation of human origins. That would have been bad enough. It was rather that he wanted to bring the supernatural into the discussion and claim that he still had science. That was anathema to scientists like Darwin. It was not just wrong—it went against everything they were trying to do as scientists, as people defining their position and important role in Victorian society. Wallace had no position and he had no role. He was indifferent to the petty shibboleths of others. If anything he thought the rules of science had for many years blocked the way to an appreciation of evolution. And so he had no hesitation at bringing in spirit forces and thinking it still qualified as science.

So come back now to the original question about rebellion. Alfred Russel Wallace was part genius and part crank. But he is not totally mystifying. His greatest work is a function of the same sorts of things that led to his most ludicrous ideas—ludicrous as judged by his contemporaries and as judged by us. The paradox is that he was a brilliant scientist and a man of strong convictions who was prepared to work for them, but he was a man outside science. Partly because of inadequate education, partly because of temperament, Wallace never (or at least rarely) played the scientific game as others were then defining it. For this reason he could accept evolution from *Vestiges* when all of the respectable scientists were pulling back. But it was also for this reason that twenty years later he could drop natural selection and argue for spirit forces when it came to humankind. Prima facie he seems the personification of inconsistency, but truly in his own funny way he was being completely consistent. The question of rebellion, however, starts to change, because this is something that goes beyond Wallace himself. In the 1840s, there was nothing particularly rebellious about Wallace's becoming an evolutionist. He belonged to a group for which his moves were almost orthodox. As Chambers shows, that group still existed in the 1860s, so one might want to say that Wallace was still no rebel, and in a way that is true. But, almost *malgré lui*, by the 1860s, Wallace had moved himself from his original group—the quasi-scientists. Thanks to his brilliant discovery of natural selection, not to mention the other work, by right and acknowledgment, he qualified to be part of the scientific establishment. He ought to have accepted—and in some respects he did accept—the rules of the game. So when Wallace invoked spirit causes, he was being rebellious. He may not have thought so, but others judged this to be, and I would argue that this is how we, too, should judge him.

ENVOI

Evolutionists know that to understand the present, we must understand the past. To understand the twentieth century, we must understand the nineteenth. Alfred Russel Wallace is a fascinating study in his own right, but his entanglements of science and (what we would judge as) nonscience show how science slowly, and at times with difficulty, broke from religion and other topics and became something in its own right. The story does not end with Wallace. Still to come, for instance, was the great growth and popularity, at the turn of the twentieth century, of vitalism—something stuck between the natural and the nonnatural, if anything is.

But do think of Wallace as a story that tells us about the century in which he died as much as about the century in which he was born.

NOTES

1. The chief sources for Wallace's life are his autobiography, *My Life: A Record of Events and Opinions* (London: Chapman and Hall, 1905) and his collected letters edited by J. Marchant, *Alfred Russel Wallace: Letters and Reminiscences* (London: Cassell and Company, Ltd., 1916). In the past decade, there has been a spate of biographies, including M. Shermer, *In Darwin's Shadow: The Life and Science of Alfred Russel Wallace* (New York: Oxford University Press, 2002); P. Raby, *Alfred Russel Wallace: A Life* (Princeton: Princeton University Press, 2001); and M. Fichman, *An Elusive Victorian: The Evolution of Alfred Russel Wallace* (Chicago: University of Chicago Press, 2004). A major problem one runs into in saying anything about Wallace is that his supporters, charmed by his gentle nature, admiring of his willingness to break all conventions when he thought he was right, convinced (without any good reason) that he is the major figure in the story of evolution and that Darwin stole all of the credit, descend like a ton of bricks at any hint that their hero's nature might be, let us say, complex. But it was, and that is the truth.

2. M. Ruse, *The Darwinian Revolution: Science Red in Tooth and Claw* (Chicago: University of Chicago Press, 1979).

3. J. Morrell and A. Thackray, *Gentlemen of Science: Early Years of the British Association for the Advancement of Science* (Oxford: Oxford University Press, 1981).

4. J. F. W. Herschel, *Preliminary Discourse on the Study of Natural Philosophy* (London: Longman, Rees, Orme, Brown, Green, and Longman, 1830).

5. W. Whewell, *The History of the Inductive Sciences*, 3 vols. (London: Parker, 1837); *The Philosophy of the Inductive Sciences*, 2 vols. (London: Parker, 1840).

6. Baden Powell, *Essays on the Spirit of the Inductive Philosophy* (London: Longman, Brown, Green, and Longmans, 1855).

7. M. Ruse, *The Evolution-Creation Struggle* (Cambridge, MA: Harvard University Press, 2005).

8. Whewell, *History of the Inductive Sciences*, 3: 588.

9. W. Cannon, "The Impact of Uniformitarianism: Two Letters from John Herschel to Charles Lyell, 1836–1837," *Proceedings of the American Philosophical Society* 105 (1961): 301–14.

10. Darwin to Asa Gray, May 22, 1860, in *The Correspondence of Charles Darwin* (Cambridge: Cambridge University Press, 1985), 8: 224.

11. A. R. Wallace, *Studies: Scientific and Social* (London: Macmillan, 1900), 2: 507.

12. M. Ruse, "Charles Darwin and Group Selection," *Annals of Science* 37 (1980): 615–30.

13. J. A. Secord, *Victorian Sensation: The Extraordinary Publication, Reception, and Secret Authorship of Vestiges of the Natural History of Creation* (Chicago: University of Chicago Press, 2000).

14. M. Ruse, *Darwinism and Its Discontents* (Cambridge: Cambridge University Press, 2006).

15. A. R. Walace, *The Geographical Distribution of Animals*, 2 vols. (London: Macmillan, 1876).

16. A. R. Wallace, "The Origin of Human Races and the Antiquity of Man Deduced from the Theory of Natural Selection," *Journal of the Anthropological Society of London* 2 (1864): clvii–clxxxvii.

17. A. R. Wallace, *Contributions to the Theory of Natural Selection: A Series of Essays* (London: Macmillan, 1870).

18. John Stuart Mill to Wallace, March 18, 1868, in Wallace, *My Life*, 2: 283.

19. Robert Chambers to Wallace, February 10, 1867, in Wallace, *My Life*, 2: 285.

PART III:
THE MIDDLE YEARS

*A*t the beginning of essay 5, I quote the eminent twentieth-century evolutionist Ernst Mayr, who rather fancied himself as a historian and philosopher of his subject. He is being rude about Darwin's contemporary and fellow English evolutionist, Herbert Spencer. To be fair to Mayr, few would think his comment wrong, let alone outrageous. Yet, there has to be something amiss in such a negative judgment. If Spencer was so silly and made so little a contribution to evolutionary thinking, why then did so many of his contemporaries take the man so seriously? Why did he outsell Darwin? Of course, being a best-seller does not guarantee excellence. Wait until later and you see what I have to say about Richard Dawkins's *The God Delusion*! But even if we do now judge Spencer's thinking of little true worth—a judgment I am inclined to make—should we not be sensitive to the possibility that others of his times (or the times when his books were still being read and taken seriously), people that we do respect today, might be influenced by Spencer? After all, Newton was much taken with alchemy, to take one example. Perhaps the notion of gravitational attraction, a shocking idea to the Cartesians, had its roots in some very weird views about occult forces. Likewise with Spencer. We might find all of his talk about homogeneity and heterogeneity a bit crazy, but others might have made use of this sort of thinking. This is just the claim I want to make about the Shifting Balance Theory (SBT) of the American population geneticist Sewall Wright. It is the basis of a revisionist interpretation I offer of twentieth-century American evolutionary biology, namely, that it owed—perhaps still owes—a major debt to Spencer, even more than to Darwin in some respects. If I am even halfway right, then (given that no one would deny the incredible significance of the SBT for American evolutionary research) this has major implications for our thinking about the truth claims of scientific theories.

The ghost of Spencer hovers above the next essay in the section. Far more than Darwin, Spencer was the person who thought that evolution could provide the foundations of a new ethics, one more suited than

Christianity for the industrialized, urban societies of the West in the late nineteenth century. Many evolutionary biologists have followed in this tradition, and my subjects are two of the key players around the middle of the century, Julian Huxley and George Gaylord Simpson. What we see in this essay is something we shall see more of in later essays, the way in which evolution was (and I believe still is) always more than just a scientific theory. It is a world picture, a secular religious world picture, with a beginning and an end, a story where we humans are not just incidentals but the apotheosis of the whole system, and as with more conventional religions it is something that offers prescriptions for moral action. Whether or not there is any truth in any of this evolutionary ethicizing, and I try to show that making a reasoned assessment is not quite as easy as one might think, we should recognize what is going on. We must do this if only because the critics of evolution are ever vigilant. We must know what parts of the evolutionary story we must defend at all costs and what parts might be more optional, more matters of taste than anything else.

The final essay in this section looks at the ways in which writers of fiction have used evolutionary themes. Expectedly, Spencer features in this essay also, particularly at the end of the nineteenth and beginning of the twentieth centuries. Starting to move down toward the present, he does begin to fade and we find writers are now picking up on more recent claims and debates in evolutionary theory. I discuss in detail a fairly recent English novel, *Enduring Love*, by Ian McEwan, which makes great use of the relatively new advances in our understanding of social behavior and of how it can be given a Darwinian causal underpinning. Yet I warn that one should be careful in assessing conclusions for they may owe less to the actual straight science than one supposes. This links with my caution expressed at the end of the previous essay, namely, that one must take care to separate what evolutionists can truly claim in the name of their science and what philosophical and other themes they and others want to weave around and through this science. One must separate what is being said by Darwinian evolutionary theory itself and what is being claimed in its name by enthusiastic Darwinians.

Five

ADAPTIVE LANDSCAPES AND DYNAMIC EQUILIBRIUM

The Spencerian Contribution to Twentieth-Century American Evolutionary Biology

> It would be quite justifiable to ignore Spencer totally in a history of
> biological ideas because his positive contributions were nil.
> (Ernst Mayr, *The Growth of Biological Thought*)

*T*he standard history of evolutionary biology—the history I myself was writing some two or more decades ago—runs somewhat like this: In the *Origin of Species*, published in 1859, Charles Darwin tried to do two things. First, he wanted to establish the fact of evolution. Second, he proposed a mechanism for evolution: natural selection brought on by a struggle for existence. Darwin was successful in his first aim. Very soon after the publication of his book, the educated world—scientists and laypeople—converted over to evolution. It became the accepted way of thinking about life's origins, including our own. Darwin was unsuccessful in his second aim. Almost no one took up natural selection as a working cause of evolutionary change—rather, a host of alternatives were preferred, including Lamarckism (the inheritance of acquired characteristics), saltationism (evolution by jumps), orthogenesis (lines of development that take on their own momentum), and more. The triumph of selection as a mechanism had to wait until the twentieth century. It was only then that biologists made the required major advances in our understanding of the mechanisms of heredity, and developed the science now known as "genetics." In the tradition of Gregor Mendel, it was realized that transmission is particulate, in the sense that the units of heredity get transmitted from generation to generation in virtually an unchanged form. Hence, there is a basis of stability on which forces for change can operate. Building on these new understandings about heredity, mathematically talented theoreticians generalized across groups and saw how natural selection can be and indeed is the major force behind long-term evolutionary change. Three people—three "population geneticists"—particularly were responsible for this advance: Ronald Fisher and J. B. S. Haldane in Britain, and Sewall Wright in America. Finally, the naturalists and experimenters put empirical flesh on the theoretical skeletons, and so the "synthetic theory of evolution" (a synthesis of Darwin and Mendel) or "neo-Darwinism" was born. The second aim of Charles Darwin was finally realized. Natural selection, the mechanism of the *Origin*, was seen as the key to evolutionary change. (This classic account can be found most explicitly in Peter Bowler's text *Evolution: The History of an Idea*.)[1]

Although a little rough and ready, the sketch above still seems to me to be a basically accurate account of the history of evolutionism in Britain. One should not neglect the fact that there were some immediate triumphs of selective explanation, notably Henry Walter Bates's analysis of butterfly mimicry[2] and (somewhat later) Raphael Weldon's analysis of crab carapace dimensions.[3] But these really were exceptions rather than the rule. For real causal movement forward, the key work after Darwin was Fisher's *The Genetical Theory of Natural Selection*, published in 1930.[4] It is true that this book had some highly nonscientific ideas lurking beneath the surface: Fisher's ardent eugenicism for a start and his commitment to Anglican Christianity for a second.[5] Nevertheless, it was a work—and was seen to be and appreciated as a work—that made natural selection *the* evolutionary mechanism first and foremost. As an account of the history of evolutionism in America, however, I argue that the above-given sketch is totally misleading. I do not want to say that, in the New World, Darwin's ideas had no influence in the first half of the twentieth century and less-than-realized in the second half—there are today many first-class, fully committed American Darwinian evolutionists (meaning evolutionists for whom natural selection is the only mechanism that really counts)—but I would say that, in the American version of this revitalized evolutionism (Mendelized evolution, that is), Charles Darwin was not the major influence. I would go further and say that there was a major influence, acknowledged or not, and this was Darwin's fellow English evolutionist, Herbert Spencer.

I would argue indeed that, notwithstanding my concessions just above, we still find significant traces of Spencerian thought in American evolutionary biology today. But in this essay I shall not attempt to make the overall case, restricting myself to one task only: showing the importance of Spencer's thinking on the evolutionary theorizing of the American contributor to the foundation of population genetics, Sewall Wright. This man, born in 1889 (and who died at the great age of ninety-eight), was trained at Harvard by the pioneering geneticist W. E. Castle, worked then for ten years at the United States Department of Agriculture, and finally went to the biology department at the University of Chicago, where in the early 1930s he published what he called the "Shifting Balance Theory of Evolution."[6] I certainly agree with the conventional history of evolutionism that this theory played a key role in the subsequent development of evolutionary research in America, most importantly through its great influence on the thinking of the Russian-born evolutionist Theodosius Dobzhansky. The latter's *Genetics and the Origin of Species* (published in 1937)[7] is rightfully acknowledged as the paradigm-creating work that led to almost everything that followed, including (paradoxically and amusingly) Ernst Mayr's wonderful work *Systematics and the Origin of Species* (published in 1942).[8] So, restricted though my aims may be, if I succeed in my task I will have done much to make the overall case at least plausible and worthy of further investigation.

COMPARING DARWIN AND SPENCER

Let us begin at the beginning. What was Charles Darwin's thinking on evolution? What was Herbert Spencer's thinking on evolution? In Darwin's case, it is best to begin with the central arguments for his mechanism of natural selection, given in early chapters of the *Origin*. First, Darwin argued for a struggle for existence:

> A struggle for existence inevitably follows from the high rate at which all organic beings tend to increase. Every being, which during its natural lifetime produces several eggs or seeds, must suffer destruction during some period of its life, and during some season or occasional year, otherwise, on the principle of geometrical increase, its numbers would quickly become so inordinately great that no country could support the product. Hence, as more individuals are produced than can possibly survive, there must in every case be a struggle for existence, either one individual with another of the same species, or with the individuals of distinct species, or with the physical conditions of life. It is the doctrine of Malthus applied with manifold force to the whole animal and vegetable kingdoms; for in this case there can be no artificial increase of food, and no prudential restraint from marriage.[9]

Second, he moved on to argue for natural selection:

> Let it be borne in mind in what an endless number of strange peculiarities our domestic productions, and, in a lesser degree, those under nature, vary; and how strong the hereditary tendency is. Under domestication, it may be truly said that the whole organization becomes in some degree plastic. Let it be borne in mind how infinitely complex and close-fitting are the mutual relations of all organic beings to each other and to their physical conditions of life. Can it, then, be thought improbable, seeing that variations useful to man have undoubtedly occurred, that other variations useful in some way to each being in the great and complex battle of life, should sometimes occur in the course of thousands of generations? If such do occur, can we doubt (remembering that many more individuals are born than can possibly survive) that individuals having any advantage, however slight, over others, would have the best chance of surviving and of procreating their kind? On the other hand we may feel sure that any variation in the least degree injurious would be rigidly destroyed. This preservation of favourable variations and the rejection of injurious variations, I call Natural Selection.[10]

Backing the direct case for selection, Darwin also argued analogically, using the evidence and techniques of artificial selection, the work of the breeders of animals and plants for profit and pleasure—bigger and shaggier sheep, fleshier turnips, stronger bulldogs, and fancier pigeons. If we humans can do so much, he claimed, nature can do far better. And then, with the case made for selection, moving now to the arguments that really convinced people of the fact of evolution, Darwin trawled through the whole spectrum of biological studies, showing how his thinking throws light on so many different and diverse areas of interest and inquiry. Instinct, paleontology, biogeographical distributions, morphology, embryology, taxonomy, and more—all of these are made reasonable by evolution through selection, and conversely evolution through selection is justified and confirmed by its explanatory successes over such a wide area. As in the best theories—astronomy, optics, geology—there was at the heart of Darwin's thinking a "consilience of inductions," as it was termed by his former mentor, the philosopher William Whewell.[11]

With an eye to future discussion, a matter of some interest is where exactly Darwin stood on the subject of progress. Was Darwin committed to a view of evolution that saw an upward rise from the primitive and simple, the monad, to the sophisticated and complex, notably man? Although there is still some debate about this, the unequivocal answer is that Darwinism equivocates! Such an upward rise was always part of Darwin's own personal view of life's history. He did indeed believe in progress. However, with respect to his science—with respect to his evolutionism—for Darwin, progress was always somewhat problematical. He did not see it as built into the process, and could see that in respects selection rather points in the other direction. There is a relativism to selection that is antithetical to progress. But Darwin wanted biological progress, and so he got it. He decided that progress does occur, and that the key functional phenomena are what today's evolutionists call "arms races"—lines compete against each other and eventually one wins, and even more eventually there is an absolute winner. In the language of today's most ardent Darwinian, Richard Dawkins, this winner is that organism with the biggest on-board computer, namely *Homo sapiens*.[12]

I am suggesting therefore that—for all his strong personal commitment to the idea—as a selectionist, Darwin felt that progress is added on, rather than built in. And this is a good point at which to introduce Herbert Spencer, for his evolutionism starts (continues and finishes) with progress. It is the very backbone of his thinking, to use an appropriate metaphor. For him, progress was not so much an empirical finding but a metaphysical presupposition of his view of history. It ran through everything, from the most primitive forms of culture to the evolution of our own species:

> Now, we propose in the first place to show, that this law of organic progress is the law of all progress. Whether it be in the development of the Earth, in the development of Life upon its surface, in the development of Society, of Government, of Manufactures, of Commerce, of Language, Literature, Science, Art, this same evolution of the simple into the complex, through successive differentiations, holds throughout. From the earliest traceable cosmical changes down to the latest results of civilization, we shall find that the transformation of the homogeneous into the heterogeneous, is that in which Progress essentially consists.[13]

Note the great importance for Spencer of what we might call the organic metaphor. He thinks hierarchically, from cell to organism, to state, to whole. He is quite explicit in thinking of the state as a kind of organism, and that progress at one level is mirrored by progress at another level. Which brings up the question of causes or mechanisms. Here, Spencer showed an eclectic synthesis of German morphology and British thermodynamics, seasoned with a good dash of British nonconformist thinking on society and the desirable underlying economic forces, arguing (again perhaps more metaphysically than empirically) that nature starts in a condition of uniformity—what he called "homogeneity"—and tends naturally to a condition of complexity—what he called "heterogeneity." Why should this be so? Apparently it follows directly from the fact that causality tends to be open ended, inasmuch as one cause leads to multiple effects, rather than many causes leading to one effect. There is always a kind of explosion or expansion outward, as the simple and uniform tends to the complex and diverse. This happens at all levels of the hierarchy—organisms, states, whatever. Something internal or external jogs or disturbs the state of being, and the multiplying causal process kicks in. More than this, however, for as the process of complexification is occurring, there is a tendency to move upward to a higher level of existence. Life—everything—is rather like the incoming tide, set on its end. There are surges forward, followed by moments or periods of consolidation, then further surges forward, with overall gain happening over and over again. Disturbance leads to the attempt to move back to a state of rest, but the new state is never that of the old state—it is more heterogeneous, and higher. Overall, therefore, evolution can be described (as it came to be known) as an exemplification of "dynamic equilibrium."[14]

Did Spencer have time for more mundane processes, like natural selection? As it happens, after Darwin himself had discovered the process, although before Darwin moved into print, Spencer wrote of selection as a contributing factor to the evolutionary process.[15] But—although it was he who provided the alternative name of the "survival of the fittest"—selection was ever a secondary mechanism for Spencer. He opted for so-called Lamarckism as primary, that is, for the inheritance of acquired characteristics. Like Darwin, Spencer thought the Malthusian explosion was important, but unlike Darwin, Spencer thought that the chief effect would be to spur organisms to greater effort, thus bringing on their evolution up the chain of being, as their simple forms transmuted into the more complex. Darwin himself, one might add, always had a place for Lamarckism, but for him it was selection first and Lamarck second. For his fellow Englishman, it was Lamarck first and selection second.

SOCIAL DARWINISM

Turn now to another important background issue. Darwin and Spencer alike thought always that their biological theorizing had implications for broader societal questions—questions about culture and about society and about gender relationships and about race and much more. One has only to read the *Descent of Man* to see how very important these issues were for Darwin. One has only to read anything by Herbert Spencer to see how very important these issues were for Spencer. To be quite frank, there is more overlap in some of the biological-cum-social thinking of Darwin and Spencer than today's Darwinians always are happy to acknowledge. Darwin, for instance, has some very Victorian views on the virtues of the class structure and on capitalism and much more.[16] But although the move from biology to society became known as "Social Darwinism," all historians properly agree that in many if not most respects it was Spencer who blazed the trail.[17] It was he who saw the processes of biology and culture, taken broadly, as being similar if not identical. It was he who saw natural selection as being the biological equivalent of the societal process of laissez faire, and who urged socioeconomic nostrums in the name of the overall metaphysical processes of life. Not that Spencer necessarily always argued from biology to society—temporally and conceptually it was often the other way. The point was that it was the big picture that counted. In what Spencer grandly called the "Synthetic Philosophy," it was evolution as a world picture that counted.

And this last point was surely a major factor in Spencer's great general success in late Victorian Britain and (even more) in America in the postbellum years. Once people were over the initial shock of the idea of transmutation—and for many this was a pretty minor shock—they not only became evolutionists but became positive enthusiasts. Indeed, as I have argued at length elsewhere, one can properly say that evolution aspired less to being a science (just as well, for in this direction at best it was second-rate) and more to being something akin to a secular metaphysics or religion.[18] It was seen as the alternative to the Christianity that had failed (or for some as the revitalizing force of the Christianity that had failed), and as something—with its story of origins and its drama of the human rise to the top—that answered all of the questions left hanging by the inadequate myths of the past. Good religions—that is to say, the monotheistic religions of the West—have social implications and promote ethical dictates about proper behavior. Love your neighbor as yourself, and so forth. Given Spencer's vigor in this direction, his positive embrace of an overall view of the evolutionary process, one that is ever moving and surging forward and that pushes humans higher and yet higher, it is hardly therefore surprising that it was he—far more than Charles Darwin—who came to epitomize the evolutionary way of thinking. Chinese intellectuals who became evolutionists became Spencerian evolutionists.[19] Indians who aspired to become intellectuals likewise became evolutionists and this meant Spencerian evolutionists.

Kim by Rudyard Kipling, by far the greatest novel written about the

Raj, has a central character, a greasy Bengali babu (clerk), who proves in the end to be the most intelligent and the bravest:

> Kim smoked slowly, revolving the business, so far as he understood it, in his quick mind.
>
> "Then thou goest forth to follow the strangers?"
>
> "No. To meet them. They are coming in to Simla to send down their horns and heads to be dressed at Calcutta. They are exclusively sporting gentlemen, and they are allowed special faceelities by the Government. Of course, we always do that. It is our British pride."
>
> "Then what is to fear from them?"
>
> "By Jove, they are not black people. I can do all sorts of things with black people, of course. They are Russians, and highly unscrupulous people. I—I do not want to consort with them without a witness."
>
> "Will they kill thee?"
>
> "Oah, thatt is nothing. I am good enough Herbert Spencerian, I trust, to meet little thing like death, which is all in my fate, you know. But—but they may beat me."[20]

It was Spencer whose books were those to which one turned first, and who was taken as the definite authority on matters of morality and custom and proper behavior—in the private and the public spheres. Nor should one think that Spencer's influence was restricted only to one segment of society, specifically to that segment that favored free economic competition—success to the winner, and widows and children to the wall. It is the first axiom of religion that true believers rarely if ever achieve uniformity of belief, especially in matters of morality—for every Quaker or Mennonite pacifist, there is a military chaplain urging one on to kill in the name of Jesus. Likewise with Social Darwinism. Businessmen liked unrestrained competition and justified this liking in the name of evolution. Bureaucrats liked organization and state control, and they, too, justified themselves in the name of evolution. Even the Marxists got on the bandwagon, for their theorizing often owed far more to

Spencer and his works than they did to Marx and *Capital*.[21] Many were the American children who received a copy of one of Herbert Spencer's works on school prize day. Many were the young men who joined a discussion group to argue over the Synthetic Philosophy. Many were the rich men who bought their ways into the kingdom of heaven through the promotion of heterogeneity over homogeneity. Many were the poor men who fought the bosses in the name of dynamic equilibrium. (Duncan gives some idea from that time about the immense influence of Spencer.[22] Russett gives a more measured historical assessment.)[23]

This being so—as we leave the nineteenth century and turn toward and into the twentieth—there is one very obvious prediction. One would expect to see in American evolutionary biology the influence of Herbert Spencer. Earlier on, in the immediate post-*Origin* period, the greatest influence was probably the vehement antievolutionist, Swiss transplant and Harvard professor Louis Agassiz.[24] Educated by the *Naturphilosophen* (with a strong dash of Georges Cuvier) Agassiz could never accept species change, but all of his students (including his own son) went over the divide. And as they went, they took with them all sorts of beliefs about archetypes and homologies and upward change and more. But even from the first, Spencer started to fuse in, hardly that surprising since it is probable that he himself drew on at least some German sources (never an easy matter to decide, given Spencer's reluctance to acknowledge the influence of others). Edward D. Cope, the great paleontologist, was open about the influence of Spencer.[25] And, more and more, the same started to be true of others as the century progressed, and as the Englishman's fame and writings spread, and as what professional evolution there was became increasingly frustrated by inadequate techniques and theory and as evolution's public status as a secular religion became yet more firmly established and acknowledged.[26]

Consider the thinking about evolution by E. G. Conklin, a major cytologist, great influence on the structure and running of American biology, and frequent writer on evolution:

> Life itself, as well as evolution, is a continual adjustment of internal to external conditions, a balance between constructive and destructive processes, a combination of differentiation and integration, of variation and inheritance, a compromise between the needs of the individual and

those of the species. And in addition to these conflicting relations we find in man the opposition of instinct and intelligence, emotion and reason, selfishness and altruism, individual freedom and social obligation. Progress is the product of the harmonious correlation of organism and environment, specialization and co-operation, instinct and intelligence, liberty and duty.[27]

"Adjustment of internal to external," "balance between constructive and destructive," "combination of differentiation and integration," "progress the product of the harmonious correlation" . . . if this is not pure Herbert Spencer, I do not know what is.

But you may interject that this is all something true of evolutionary biology while it was still in its prescientific (or second-class scientific) state, before it matured thanks to the efforts of the great theoretical population geneticists. Once Fisher, Haldane, and (especially in our case) Sewall Wright had set to work, evolution as a science was moved right up—and one of the major features of this move up is that Spencerianism must have been expelled. From then on (say around 1930), it was pure Darwinism all of the way. No one would deny Spencer at the beginning of the century, just as no one should deny Darwin before the middle of the century. Yet is this really true? I should say (by way of warning) that proving a positive case—that Spencer was indeed an important influence—cannot be an easy task. Along with the intellectual maturing, which came with the arrival of population genetics, came the social drive to professionalize evolutionary theory. No longer was evolution to be something functioning primarily as a secular substitute for Christianity, confined (as it then was) to museums and to the popular lecture hall and to the magazine for the general reader. It was to be part of real science, in universities, done by trained specialists, getting students and grants. But one of the chief marks of professional science is that it stays strictly away from religion and "philosophy," where something like Social Darwinism—with its overt moral and political agenda—is precisely what is meant and feared. Hence, even if Spencer was an influence, from about 1930 on—given especially that Social Darwinism was now being tied into some of the worst excrescences of the twentieth century (just look at *Mein Kampf* for starters)—one should expect a reluctance to parade the fact too obviously. Given that Ernst Mayr was one of those who worked longest and hardest to upgrade evolutionary biology, it is little wonder that the history he wrote was one that included the sentiment expressed at the head of this essay.

THE SHIFTING BALANCE THEORY AND ITS DISCONTENTS

But enough of pouring water on the altar. Can it be lit? Even though the time that Mendel was coming into evolutionary biology was also just the time when evolutionists would be eager to deny or downplay the influence of Spencer, can we nevertheless establish the continued influence of Spencer? Specifically, can we establish the continued influence of Spencer on and in Sewall Wright's Shifting Balance Theory of evolution? To start the ball rolling, let us begin by noting a number of interesting (and I shall argue significant) facts about this theory.

First, no one could follow the theory. True, this is a bit of an exaggeration. But not that much. The theory is presented in two places. First, in a long and rather technical paper in *Genetics* in 1931. Second, in a much shorter poster paper—a *Reader's Digest*-type condensation—given at an international congress of genetics in 1932. It is the first paper that lays out the guts of the theory.[28] It is the second paper that people read and thought they understood.[29] No one could follow the math of the first paper. Dobzhansky was open that Wright's calculations were beyond him.[30] When later he and Wright coauthored papers, Dobzhansky admitted almost proudly that he understood the first and last lines only. And the same blankness toward theory was true of the other important evolutionists of the period. Mayr was never able to follow one symbol next to another. Again and again he declaimed against formal techniques and products.[31] And others were little better. Some years ago (May 25, 1988), I interviewed the botanist of the group, G. Ledyard Stebbins, author of *Variation and Evolution in Plants*.[32] He saw the Wright paper at the congress and was at once very excited (or, if memory was improving on the occasion, certainly became very excited). In the same interview, Stebbins admitted without hesitation or attempt to conceal that the reason for the excitement was that here was something he could understand. No math.

Second, the theory—especially the theory as given in the second paper—was seriously confused. It was really seriously confused to the point of incoherence. The key notion of this second paper was that of an adaptive landscape. This metaphor enabled Wright to collapse down a huge amount of information into an easily graspable, visual picture. A picture supposed to be in three dimensions, with x and y axes showing where one finds organisms (or are they groups?), and the z axis (generally not shown in a two-dimensional picture) sticking out and representing fitness. The higher up the landscape, the fitter the inhabitant. Supposedly, organisms sit on tops of peaks, basically kept up there by selection. Every now and then however a population wanders off the top, down the side, and if lucky (most are not) up the side of another mountain or hill. How does this happen? Through the random effects of breeding, where small-population contingencies outweigh selection—in other words, through so-called genetic drift. Again, supposedly, the reason why a population might suddenly start moving up, after a drift-driven downward journey, is that a new combination of features might get together and this (or these) would prove adaptively advantageous. Finally, some organism or group that was much better than others on nearby peaks would either beat everyone else, or its fancy new features would get spread around. The process is over until the next time.

Fine and dandy, except—as Will Provine, Wright's dedicated and splendid biographer, showed—there is really radical confusion about those x and y axes.[33] What are we actually plotting? Is it gene frequencies? If so, which genes and how and why? Can one simply split everything apart in this kind of reductionistic fashion, treating the genes like (to use Ernst Mayr's metaphor) beans in a bag? Or is it all a matter of individual genotypes, so that points on the graph represent individual organisms? In which case, why can one assume a smooth transition from one genotype to another? As Provine pointed out, the trouble is that (to get his adaptive landscape) Wright is indeed collapsing a huge amount of information into two or three dimensions. The virtue is that a lot of information can now be presented very simply. The mis-virtue is that, not only is a lot of information lost—a lot of information is confused. Wright himself had apparently never even thought about these issues until he was flagged to them. But even he had to agree that, when you start to peer into what the adaptive landscape is all about, it is a conceptual mess.

Third, the theory is false from beginning to end. It has virtually no connection at all with the real world. Recent analyses of the theory—theoretical and empirical—show that it just does not work or do the job required of it, namely explain change in an adaptive or otherwise fashion. Properly characterizing the view of Fisher as involving a selective force working on large populations, being driven to adaptive excellence, the most severe critics write as follows:

> Although the mathematics of the shifting balance theory (henceforth SBT) is complicated, its essence is simple. Wright proposed that adaptation involved a shifting balance between evolutionary forces, resulting in a three-phase process:
>
> Phase I: Genetic drift causes local populations (demes) to temporarily lose fitness, shifting across "adaptive valleys" toward new "adaptive peaks."
>
> Phase II: Selection within demes places them atop these new peaks.
>
> Phase III: Different adaptive peaks compete with each other, causing fitter peaks to spread through the entire species. (Wright believed that populations occupying higher adaptive peaks would send out more migrants, ultimately driving other populations to the highest peak.)
>
> There is thus a clear distinction between the Fisherian and Wrightian views of evolution: the former requires only that populations be larger than the reciprocal of the selective coefficient acting on a genotype, and the latter requires subdivided populations, particular forms of epistasis, genetic drift that counteracts selection, and differential migration between populations based on their genetic constitution.[34]

They write then:

> We begin our analysis with an examination of the theory itself and then discuss the data offered in its support by Wright and others. We will conclude that (1) many of Wright's motivations for the SBT were based on the problems he perceived with the alternative process of mass selection, but these problems are largely illusory; (2) although, as Wright postulated, alternative adaptive peaks separated by adaptive valleys

clearly exist, there is little evidence for the assumption that movement between peaks involves a temporary loss of fitness; (3) although phases I and II of the theory may be at least theoretically plausible, there is little theoretical support for phase III of the shifting balance, in which adaptations spread from particular populations to the entire species; (4) the few possible examples of the SB process do not increase adaptation in the way envisioned by Wright; (5) there are almost no empirical observations that are better explained by Wright's mechanism than by mass selection; and (6) because of the complexity of the SBT, it is impossible to test Wright's claim that it is a common evolutionary process. In view of these problems, we think that it is unreasonable to consider the SBT an important explanation of adaptation in nature.[35]

Now, of course, nothing thus far proves the influence of Herbert Spencer. But, even if we assume that only some of the above is true, we are forced toward an asymmetrical position. If there were good reason to think that Wright's theory was conceptually clear and essentially true, then one might simply argue that—no matter how similar it seems to anything written by anyone else, Spencer or otherwise—the reason for its taking the form that it does is that it corresponds to the way that the world is. I do not need Spencer or anyone else to tell me that I am writing in English and that this sentence has a main verb. Why then pin an influence on Spencer? At least, one cannot say that there must be one. But since Wright's theory so clearly does not correspond to physical reality, then one is driven to a search for sources. A photograph of Iowa cornfields is one thing. When Van Gogh gets out his palette and brush, it is quite another. There had to be something that made Sewall Wright come up with what he did, and the way that the world turns is not it. Let us therefore turn our gaze backwards, and do the most obvious thing. What kind of theory does Wright's look like? If you were looking for influences, what does it remind you of? The US Supreme Court building looks like a Greek building rather than a Mexican building, so let us start with that. Similarly, for the Shifting Balance Theory.

NATURAL SELECTION OR DYNAMIC EQUILIBRIUM?

Since the general consensus is that it was Darwinism redux that happened at the beginning of the 1930s, let us first ask if Wright's theory looks like the theory of the *Origin of Species*. To which only one reply is possible: You must be kidding! Landscapes, peaks, drifting down, regrouping in new formations away from selection—the point at which the really important innovations are occurring—and then and only then selection in a backup, clean-up role, with groups within the whole population fighting it out. This may be many things, but it is not Charles Darwin. Fisher, a fanatical Darwinian, saw that right off. No one could make genetic drift as significant as did Wright, and still be a Darwinian—especially since drift was not given a minor role, but played *the* crucial part in evolutionary advance. More than this: You may object—and it is certainly true—that although selection had a minor role in the early 1930s, at the end of the decade the empiricists (Dobzhansky particularly) found strong evidence of selection where once drift had been supposed. Hence, even though the early version of the theory was not very Darwinian, in potential it was Darwinian, for it could be modified and selection given a bigger role. But in a way this backfires, even though—precisely though—Wright did bring in more selection to his theory. He himself was not that bothered about the change—more accurately, he himself was supremely indifferent to the change—because for him the details of the mechanism were simply not that important. What counted was the overall picture. Think for a moment about the name of Wright's theory, something that often puzzles people: the Shifting Balance Theory of evolution. What is balanced? What is shifting? Wright is explicit on this. We have a balance—might one say an equilibrium?—and, then for various imposed reasons, this gets destabilized. Then we get a move—a shift—to a new position. What is balanced? They are the forces tending to similarity and those tending to difference. Those forces making for genetic homogeneity and those making for genetic heterogeneity. This is Wright's language, not mine:

> Evolution as a process of cumulative change depends on a proper balance of the conditions, which, at each level of organization—gene, chromosome, cell individual, local race—make for genetic homo-

geneity or genetic heterogeneity of the species. . . . The type and rate
of evolution in such a system depend on the balance among the evolutionary pressures considered here.[36]

By this time—and if nothing else is twigging you, the hierarchical language should—the case is almost overwhelming that with the Shifting
Balance Theory we are looking, not at something Darwinian, but at
something very Spencerian. We have stability, we have disruption, we
have a vital non-Darwinian shift that creates new innovations, we have a
return to stability. We have, to use Wright's own language, a tension
between "homogeneity" and "heterogeneity." We have, to use a phrase,
dynamic equilibrium. Of course the two positions—Wright's and
Spencer's—are not identical. Wright has genes. Spencer does not.
Spencer is a Lamarckian. Wright is not. But without denying these differences, in a sense they are trivial, because for both men what really
counts is the big picture. The details of the mechanisms can be filled in
around this picture. Both are prepared to use selection, but for neither is
it the be-all and end-all—as it was for Darwin and Fisher.

But what about the all-important question of progress? Does not the
case for similarity come tumbling right down here? Spencer was the ultimate progressionist. Wright has not a mention of it in his theory. This
surely divides them. It is at this point that we have to go back to the
matter of professionalism. Genetics was an insecure subject at the beginning of the last century. For ten years, Wright was in the USDA—and
(speaking now with the authority of thirty-five years of teaching at an
agricultural college) everyone knows that, in the academic pecking order,
agriculture rates just above education, and even below sociology. Then,
when he got a faculty position, Wright was low on the status totem pole.
Amazing to us today, back in those days, genetics came below embryology. And, evolution was even more insecure. It simply had to be presented without a whiff of philosophy or religion or whatever. Wright
knew that and admitted that, even in his papers. He had to stay away
from "speculation." And right at the top of the maxi-to-be-avoideds
would be speculation about progress, and the triumph of humans, the
Anglo-Saxon humans in particular. But it does not mean that progress
was not there in Wright's work. It does not mean that the progress was
not there, in an absolutely fundamental Spencerian fashion (rather than

a Darwinian add-on fashion). For a progressionism booster, who was also desperate to be seen as a professionalism booster, the best kind of theory would be one that had no necessary progress built in, but that would lend itself very readily to a progress-impregnated interpretation. That would beg for a progress-impregnated interpretation. And this of course was precisely the Shifting Balance Theory! The landscape could be like a water bed. As one peak goes up another goes down, and ultimately you are right back where you started. Or it could be like the Himalayas, where things are pretty much fixed in rock, and Mount Everest is not about to sink below its neighbors. So no one could accuse Wright of being a progressionist with respect to the theory.

But progress was there for the taking if you wanted it. And Wright wanted it. Progress is right there in the first little pre-paper sketch that Wright sent to Fisher. And it is there from then on. Wright had some very strange metaphysical beliefs about everything—everything!—having consciousness, from molecules to men, and that perhaps we are on the way up to a kind of superorganism, with superconsciousness. This "panpsychic monism" is deeply progressionist, hierarchically and temporally:

> The greatest difficulty is in appreciating the possibility of the integration of many largely isolated minds into a higher unitary field of consciousness such as must necessarily occur under this viewpoint in the organism in relation to its cells; in these in relation to their molecules and in these in relation to their molecules and these in relation to more ultimate entities. The observable hierarchy of physical organization must be the external aspect of a hierarchy of mind.[37]

So what I am concluding is that with respect to progress, as with respect to much else, Wright's is just the kind of theory that a 1930s Spencerian evolutionist, with aspirations to professionalism, would be expected to produce.

Let me add a couple of historical footnotes by way of backing. First, everyone after Wright—from Dobzhansky on—interpreted the adaptive landscape scenario in a progressivist fashion. Mayr, Stebbins, G. G. Simpson (author of *Tempo and Mode in Evolution* and the paleontologist of the group) were all ardent progressionists and took Wright's theory as their starting point. So the progressionism is a figment of more imaginations than mine. Second, in line with his indifference to the rising flood of selectionism, Wright basically was not that bothered about evolution as a science after he had published his theory. It is true that, in the late 1930s and early 1940s, Wright wrote some fundamental papers with Dobzhansky, but the impetus came from Dobzhansky, and Wright got out of the collaboration as soon as he could. He never worked on evolution himself, devoting his energies to increasingly dated genetic studies with guinea pigs. And he supervised just a couple of evolutionary theses out of over thirty in all, and simply did not want to talk about evolution with his students. But he loved to talk about it with philosophers and theologians, especially his pals at the University of Chicago Faculty Club. This is really odd at the best of times, and it is truly odd without the missing factor of Spencer's influence.[38]

SOURCES

We come to the final part of the discussion on Wright. Is there any direct evidence of a Spencerian input? I am certainly not going to say that Spencer was the only input. Apart from all of the genetics, Provine has shown (what Wright himself acknowledged) that the time at the USDA was crucially important. In particular, Wright did a massive analytic study of shorthorn cattle, and this convinced him that selection could not work in large groups—the secret is breaking the population into small isolated numbers, trying to effect change first in them, and only later returning to the large group. This was undoubtedly built into the Shifting Balance Theory. But this in itself did not make for a theory of evolution, and so we start to look farther afield. One influence that Wright acknowledged was the French philosopher Henri Bergson, author of the vitalist classic *Creative Evolution*, and one can certainly see traces of this. Bergson was no great enthusiast for selection, thinking that it did not solve the problem of new and innovative features—the very things that Wright highlighted as needing more than mere selection. More than this: There is in Bergson a hint of the adaptive landscape metaphor.[39] And Bergson was an ardent progressionist—although, showing that we should not take Bergson as squeezing out Spencer, we should note that Bergson himself always acknowledged the importance of Spencer in his own thinking, as something that directed his own thought on evolution.

Another influence on Wright apparently was a now-unknown chemist from Liverpool University in England, one Benjamin Moore. He wrote in terms that seem almost to be cribbed from the Synthetic Philosophy:

> It is only necessary for the atomic basis to our chemistry to realize that the atom, just like the chemical molecule at the different stage, or the fixed organic species of the biologist, is a point of the stable equilibrium in upward evolution. Between each two such points there lies a region of unstable equilibrium, and as matter becomes more charged with energy, surging and transformations occur, and in the greater number of cases when the cycle is complete, the matter drops back again to its stable point. But occasionally when a supply of energy at high-potential, or concentration, is available, there is a huge wave of uplifting which carries the matter involved over a hill crest into a higher hollow of stable equilibrium, and a new type of matter becomes evolved at the expense of kinetic energy passing over into latent energy or potentia.[40]

I should say that I find no evidence whatsoever that one potential influence, American Pragmatism, played any role at all. That kind of thinking seems to have had no appeal whatsoever for Wright. So finally, we start to corner in on Spencer himself. There is the home and early background. Sewall's father was an economist who taught his own son as an undergraduate, and who later was on the Harvard faculty when the son was a graduate student, and who apparently was much given to progressivist-type thinking in social and other spheres. Sewall's first teacher of biology, Wilhelmine Marie Entemann (Key) was apparently a Spencer enthusiast and had been herself educated by Spencer followers—her doctoral work was supervised by Charles Otis Whitman, an explicit enthusiast for dynamic equilibrium thinking. Sewall and his brother Quincy—the latter a specialist in international law and also to become (like Sewall) a professor at Chicago—corresponded knowledgeably about Spencer, and Sewall (as a student) apparently had a picture of Spencer (Darwin also) on his wall.[41] And then, above all, there was the influence at Harvard of the chemist (and Sewall's teacher) L. J. Henderson, author of the well-known work *The Fitness of the Environment*, and Spencer fanatic.

Through Henderson's writings there are all sorts of organismic analogies, movements upward, changes from simplicity to complexity, and most prominently, that ever-changing flow to and from a state of balance. As Henderson said explicitly, "Spencer's belief in the tendency toward dynamic equilibrium in all things is of course fully justified."[42]

The student was brought under the spell—"I was always very much impressed with Henderson's ideas"[43]—and acknowledged explicitly the direct influence back to Spencer. "I found him a very stimulating lecturer and got lots of ideas from him, 'condition of dynamic equilibrium' etc."[44] And the young thinker worked things out, particularly in letters to Quincy. The organismic analogy:

> Thus the body is not an absolute monarchy in which the bulk of the cells are mere mechanisms, directed in every action by a central unit. It is democracy or perhaps better is limited monarchy. In the main each part knows what to do and does it of its own accord, as occasion arises. Regulation from outside comes rather from suggestions from numerous peers, not in a single command from above.[45]

The hierarchical thinking, linking evolution and equilibrium:

> My original idea was to classify all sciences by the unit of organization—electron, atom, animal, etc.—with which they deal subdividing on a fourfold basis—
>
> A. Condition in equilibrium
>   1. Description of organization
>   2. Mechanism of maintenance of equil.
>
> B. Change of equilibrium (Evolution)
>   3. Description of changes (history)
>   4. Mechanism of change[46]

And another attempt, again linking evolution and equilibrium:

> The difficulty of classification is well illustrated by my own science, genetics—from one point of view it deals with the organization of the cell and has very close relations with cytology, then it deals with the mechanism of individual development—the mode in which develop-

mental factors are represented in the one cell stage,—and finally it deals with both the maintenance of equilibrium in the species (heredity) but also the mechanisms of change in this equilibrium (variation by recombination of factors and otherwise).[47]

And wrapping everything up in terms of progress:

> Darwinists would hold that the most rapid evolution would follow from a happy mean between conditions which permit the existence of a wide range of variations, many of them more or less injurious—which can recombine in all possible ways—and conditions which tend to eliminate the more unfit. To use a human analogy, we do not expect civilization to advance most rapidly either in the arctic zone where existence depends on following one very definite mode of life or in the tropics where conditions of life are too easy. . . . The greatest progress should result in a society which is neither crystallized into a caste system nor so fluid that individuals of a family, which has produced favorable variations and done much for progress in the past, receive no advantage over inferior families. The problem of statesmanship is to adjust laws so that there is just the degree of viscosity in all respects which gives the maximum progress. It is a problem of maxima and minima and therefore much more difficult than progress toward an absolute democratic or absolute aristocratic ideal.[48]

If this does not all add up to a smoking gun, I do not know what does. The Shifting Balance Theory—the genesis of which, incidentally, apparently goes back long before 1930 (apparently it was first written up around the time Wright went to Chicago, that is, around 1925)—was Herbert Spencer updated. There are links back through Henderson and his influence to the Synthetic Theorist himself. R. A. Fisher was a Darwinian. Sewall Wright was not.

CONCLUSION

True to my promise, I will leave things here. I will forbear mentioning that the Stephen Jay Gould/Niles Eldredge non-Darwinian, paleontological theory of Punctuated Equilibrium sounds very much like a Spencerian offshoot to me (especially in Gould's final massive testament,

The Structure of Evolutionary Theory). I will leave unsaid the fact that the arch-progressionist of American evolutionary biology today, Edward O. Wilson, has on his wall a picture of Herbert Spencer, whom he much admires. ("Great man, Mike! Great man!") Debunking one myth per paper is enough heresy even for me.

NOTES

1. P. Bowler, *Evolution: The History of an Idea* (Berkeley: University of California Press, 1984).

2. H. W. Bates, "Contributions to an Insect Fauna of the Amazon Valley," *Transactions of the Linnean Society of London*, 1862.

3. W. F. R. Weldon, "Presidential Address to the Zoological Section of the British Association," *Transactions of the British Association* (1898): 887–902.

4. R. A. Fisher, *The Genetical Theory of Natural Selection* (Oxford: Oxford University Press, 1930).

5. M. Ruse, *Monad to Man: The Concept of Progress in Evolutionary Biology* (Cambridge, MA: Harvard University Press, 1996).

6. W. B. Provine, *Sewall Wright and Evolutionary Biology* (Chicago: University of Chicago Press, 1986).

7. T. Dobzhansky, *Genetics and the Origin of Species* (New York: Columbia University Press, 1937).

8. E. Mayr, *Systematics and the Origin of Species* (New York: Columbia University Press, 1942).

9. C. Darwin, *On the Origin of Species by Means of Natural Selection, or the Preservation of Favoured Races in the Struggle for Life* (London: John Murray, 1859), p. 63.

10. Ibid., pp. 80–81.

11. W. Whewell, *The Philosophy of the Inductive Sciences*, 2 vols. (London: Parker, 1840).

12. R. Dawkins, *The Blind Watchmaker* (New York: Norton, 1986).

13. H. Spencer, "Progress: Its Law and Cause," *Westminster Review* LXVII (1857): 244.

14. H. Spencer, *First Principles* (London: Williams and Norgate, 1862).

15. H. Spencer, "A Theory of Population, Deduced from the General Law of Animal Fertility," *Westminster Review* 1 (1852): 468–501.

16. M. Ruse, *The Darwinian Revolution: Science Red in Tooth and Claw* (Chicago: University of Chicago Press, 1979).

17. R. J. Richards, *Darwin and the Emergence of Evolutionary Theories of Mind and Behavior* (Chicago: University of Chicago Press, 1987).

18. M. Ruse, *The Evolution-Creation Struggle* (Cambridge, MA: Harvard University Press, 2005).

19. J. R. Pusey, *China and Charles Darwin* (Cambridge, MA: Harvard University Press, 1983).

20. R. Kipling, *Kim* (Garden City, NY: Doubleday, Page & Co., 1914), p. 356.

21. M. Pittenger, *American Socialists and Evolutionary Thought, 1870–1920* (Madison: University of Wisconsin Press, 1993).

22. D. Duncan, ed., *Life and Letters of Herbert Spencer* (London: Williams and Norgate, 1908).

23. C. E. Russett, *The Concept of Equilibrium in American Social Thought* (New Haven, CT: Yale University Press, 1966).

24. E. Lurie, *Louis Agassiz: A Life in Science* (Chicago: Chicago University Press, 1960).

25. H. F. Osborn, *Cope: Master Naturalist: The Life and Writings of Edward Drinker Cope* (Princeton, NJ: Princeton University Press, 1931).

26. Ruse, *Monad to Man*.

27. E. G. Conklin, *The Direction of Human Evolution* (London: Oxford University Press, 1921), p. 87.

28. S. Wright, "Evolution in Mendelian Populations," *Genetics* 16 (1931): 97–159.

29. S. Wright, "The Roles of Mutation, Inbreeding, Crossbreeding and Selection in Evolution," *Proceedings of the Sixth International Congress of Genetics* 1 (1932): 356–66.

30. E. Mayr and W. Provine, *The Evolutionary Synthesis: Perspectives on the Unification of Biology* (Cambridge, MA: Harvard University Press, 1980).

31. E. Mayr, *Towards a New Philosophy of Biology: Observations of an Evolutionist* (Cambridge, MA: Belknap, 1988).

32. G. L. Stebbins, *Variation and Evolution in Plants* (New York: Columbia University Press, 1950).

33. Provine, *Sewall Wright and Evolutionary Biology*.

34. J. A. Coyne, N. H. Barton, and M. Turelli, "Perspective: A Critique of Sewall Wright's Shifting Balance Theory of Evolution," *Evolution* 51, no. 3 (1997): 643–44.

35. Ibid., pp. 644–45.

36. Wright, "Evolution in Mendelian Populations," p. 158.

37. Wright to J. T. McNeill, November 12, 1943, in Sewall Wright Papers, American Philosophical Society, Philadelphia.

38. I discuss this in *Monad to Man*. Two key pieces of information were letters to me, one from Janice Spofford, Wright's last doctoral student, August 10, 1995, and the other from the distinguished population geneticist and long-time friend and colleague of Wright after he retired from Chicago and went for over thirty years (in the event, longer than his time as a full-time faculty member at Chicago!) to Wisconsin, James Crow, August 14, 1995.

39. Jean Gayon, in his important *Darwin et L'après Darwin: Une histoire de l'hypothèse de sélection naturelle* (Paris: Kimé, 1992), suggests that Wright might have been influenced by a diagram in a 1931 paper by Haldane, which was drawn by C. H. Waddington.

40. B. Moore, *The Origin and Nature of Life* (London: Williams and Norgate, 1913), p. 40.

41. Information from Sewall to Quincy Wright, December 14, 1915. These and the subsequent letters are in the Quincy Wright Papers, University of Chicago. In 1925 Sewall joined Quincy (a political scientist and important theorist on war) on the Chicago faculty and at that point the letters stop.

42. L. J. Henderson, *The Order of Nature* (Cambridge, MA: Harvard University Press, 1917), p. 138.

43. Interview with Provine, June 4, 1976, S. Wright Papers.

44. Letter to Quincy Wright, January 10, 1916.

45. Letter December 14, 1915.

46. Letter February 27, 1916.

47. Ibid.

48. Letter October 17, 1915.

Six

JULIAN HUXLEY AND GEORGE GAYLORD SIMPSON ON EVOLUTION AND ETHICS

If Julian Huxley is the Herbert Spencer of twentieth-century Darwinism, George Gaylord Simpson may in some ways be considered its Thomas Henry Huxley.
    —John C. Greene, *Science, Ideology and World View*

hilosophers usually think that evolutionary ethics—the attempt to locate and ground morality in our biological origins—met its Waterloo in the crucial year of 1903. It was then that the English philosopher G. E. Moore published his devastating critique of the ideas of the prominent nineteenth-century evolutionary ethicist Herbert Spencer. In a definitive manner, Moore's *Principia Ethica* showed that Spencer and all who think like him are guilty of that gross conceptual mistake that Moore labeled "the naturalistic fallacy." Before Moore, thinks the philosopher, evolutionary ethics flourished like the rank weed that it is. After Moore, thinks the philosopher, evolutionary ethics lay smoldering on the bonfire of discarded ideas. And a good thing too, for there have been few excesses of nineteenth-century capitalism or twentieth-century militarism and fascism that have not had biological partisans. Choose your vileness, and there has been someone prepared to defend it in the name of evolution.

Those whose inquiries have taken them beyond philosophical folklore will know that there is now little need to spend too much time on this latter charge. For years, historians have been looking at the claims of the evolutionary ethicists—"social Darwinians" as they are often called—and it is clear that, although indeed some pretty dreadful things were suggested and perhaps even done in the name of evolution, the picture was by no means uniformly black. To the contrary: some thoroughly admirable ends were promoted under the same name. Even Herbert Spencer has much to commend him, for instance in his ardent and steadfast opposition to the militarism that engulfed Europe toward the end of the nineteenth century.[1]

But what of the former claim about the force and significance of *Principia Ethica*? Here, too, the historian will point out that philosophical tradition has but a tenuous connection to reality. One does not have to deny the power of Moore's vigorous critique—although one might properly note that he was less than generous in failing to acknowledge that, a century and a half previously, David Hume had made some of the same crucial points—to acknowledge that before Moore there were very effective critics of evolutionary ethics and that after Moore there were very energetic supporters.[2] Among the former were the philosopher Henry Sidgwick[3] and the biologist Thomas Henry Huxley.[4] Among the latter was just about every evolutionist of note in the first half of this century, and there have been many more since then. They do not share one voice, but they are united in the conviction that one can and must bring evolution and ethics together in one harmonious and fruitful relationship. It simply has to matter that we humans are, as Huxley once wrote to a friend, modified monkeys and not modified dirt.[5]

Whatever the merits of the arguments for or against, to those knowledgeable of the history of evolutionary theorizing, there can be no surprise that Moore's work had so little effect and that people have gone on connecting origins and ethics. For all that the past eighty years have seen a major move toward professional science, evolution has always been more than just a scientific theory—it has ever been a philosophy, a metaphysics, a Weltanschauung, a secular religion (not so secular at times), even indeed an eschatology. And naturally, therefore, evolutionists continued to turn their attentions to morality, its nature and foundations, whether intending to support or supplement the already-existing ideas and practices, sacred or secular, or whether hoping to start afresh, in altogether new directions. In this dreadful era of war, pestilence, famine, overpopulation, pollution, the wonder would be that *Principia Ethica* had any lasting effect whatsoever—except perhaps among the philosophers, who needed little persuasion anyway.

In this discussion, I want to pick out for critical examination two of the most articulate and interesting of the evolutionary ethicists of this century: the Englishman Julian Huxley and the American George Gaylord Simpson. I should explain at once that I write now both as a historian and as a philosopher. Qua historian, Huxley and Simpson make a good contrast because, although they were good friends and correspondents on these matters, they held very different views on the nature of the evolutionary process and the ethics to which it gives rise. Nor were these idiosyncratically different views. The quote at the beginning of this essay from the pen of the doyen of historians of evolutionary thought, John Greene, supports my suspicion that Huxley and Simpson are worth considering, not just in their own right, but as representatives of two separate traditions in evolutionary ethical theorizing.

Qua philosopher, I pick out Huxley and Simpson because today there is renewed interest in evolutionary ethics, and even philosophers (now that we are over a century after *Principia Ethica*) are starting to think that perhaps our simian ancestry does count. Convinced, as an evolutionist, that the best way to understand the present is to understand the past, I use Huxley and Simpson to this end. To lay my cards openly on the table, I believe—and I am neither alone nor original in this—that of the two traditions in evolutionary ethicizing, whereas one is fruitful and forward looking, the other is not. My fellow philosophers have been right to be suspicious, but their skepticism turned them away too early. I am not going to say that either Huxley or Simpson was completely right or completely wrong (or that either would have agreed with me on what constitutes "completely right" or "completely wrong"!) but I do say that, by looking at the pertinent ideas of these two men, we shall carry discussion forward in a fruitful manner.

JULIAN SORREL HUXLEY (1887–1975)

Intellectually, Julian Huxley was born with a silver spoon in his mouth. Grandson of Thomas Henry Huxley (and brother of Aldous), he was also the great-grandson of Dr. Thomas Arnold of Rugby School (and grand-nephew of Matthew). Educated at Eton and Oxford, he worked briefly at the Rice Institute (now University) in Houston, Texas, then back at Oxford and after at Kings College, London. Soon, he turned from the straight academic life, putting in spells as secretary of the London Zoo and as the first director general of UNESCO. Increasingly his time, however, was spent as a writer of articles and books—popular and learned—as well as in being a general pundit (first on radio, later on television) on matters scientific.[6]

Huxley did creditable work on animal behavior (ethology) as well as embryology (especially on problems of comparative growth)[7]; but his great importance as a scientist was as a synthesizer, most particularly of evolutionary theory. He was one of the founders of the so-called synthetic theory of evolution, that coming together in the 1930s and 1940s of the selectionist ideas of Charles Darwin and the particulate theory of heredity that dates back to the Moravian monk Gregor Mendel. Huxley's 1942 *Evolution: The Modern Synthesis* was a landmark, as he surveyed and integrated the field as then known, covering the range of evolutionary topics, from biogeography to taxonomy, from anatomy to paleontology.

But even at his most scientific, there was always more to Huxley's work than a simple quest for empirical understanding of the world. Although, ontologically, he was at the atheistic end of the spectrum, Huxley was ever trying to make sense of life in a deeper, more spiritual sense. For him, the key concept—that which flooded his whole being and infused all of his work—was that of *progress*, the belief or doctrine that all is in a state of flux or change, and that the direction of this change is from lesser to greater, from less to more, from value-free to value-loaded or worthy or improved.

Of course, progress was not a new notion with Huxley. At a general cultural level, it was the ideology of the Enlightenment, and in biology it was the very mark of evolutionary theorizing from its eighteenth-century beginnings. The key question had always been, not whether biological progress had ever occurred, but what form it generally took. Much ink was spilled in defenses of candidates for the true criterion: complexity, intelligence, flexibility, and more. Huxley liked the idea of complexity— "High types *are* on the whole more complex than low"[8]—but what really excited him were the notions of *control* and *independence*. In his opinion, the more the organism is capable of exercising control over its particular environment and the greater its independence of this environment, the higher is such an organism up the scale of improvement. Mammals, for instance, with their methods of maintaining a constant body temperature are more in control and more independent than reptiles without such methods—and they are clearly higher up the scale.

Huxley saw humans right at the pinnacle of being. More than any other organism, they have "increased control over and independence of the environment," or in other words, they have raised "the upper level of all-round functional efficiency and of harmony of internal adjustment."[9] Seeing progress in the evolution of organisms up to our own species and, believing progress to be a social phenomenon also, Huxley also saw social progress as being that which has taken up from biological progress. "True human progress consists in increases of aesthetic, intellectual, and spiritual experience and satisfaction."[10]

Apparently, good though we are, there are further possibilities ahead:

> The Ant herself cannot philosophize–
>> While Man does that, and sees, and keeps a wife,
> And flies, and talks, and is extremely wise . . .
>> Yet our Philosophy to later Life
> Will seem but crudeness of the planet's youth,
>> Our Wisdom but a parasite of Truth.[11]

(This is from a sonnet titled "Man the Philosophizer." That this should come from the same family that two generations before had produced "Dover Beach" surely proves that not all change is progress.)

Huxley's original belief in progress probably had several sources; but an early reading of Henri Bergson's *Creative Evolution* was crucial. In a book whose preface stated candidly that "it will easily be seen how much I owe to M. Bergson," Huxley wrote:

> Civilized man is the most independent, in our [i.e., Bergson's] sense, of any animals: this he owes partly to his comparatively large size, more to

his purely mechanical complexity of body and brain, giving him the possibility of many precise and separate actions, and most to the unique machinery of part of his brain which enables him to use his size and the smoothly-working machine-actions of his body in the most varied way.[12]

Indeed, there is good evidence that (like Bergson) Huxley was attracted to some form of vitalism, believing in spirit forces driving organisms up the chain of being. Certainly, he had mystical experiences that inclined him this way:

When I was last in New York, I went for a walk, leaving Fifth Avenue and the Business section behind me, into the crowded streets near the Bowery. And while I was there, I had a sudden feeling of relief and confidence. There was Bergson's élan vital—there was assimilation causing life to exert as much pressure, though embodied here in the shape of men, as it had ever done in the earliest year of evolution:—there was the driving force of progress.[13]

However, unlike Bergson, Huxley realized that naked vitalism could never be genuine science, and so in a way his whole career can be seen as an attempt to put a scientific gloss on the philosophy—a gloss that (by and large) Huxley thought could come from Darwinian selection working on Mendelian-caused heritable organic characteristics, thus yielding ever-new adaptation of a progressive kind. "Bergson's *élan vital* can serve as a symbolic description of the thrust of life during its evolution, but not as a scientific explanation. To read *L'Évolution Créatice* is to realize that Bergson was a writer of great vision but with little biological understanding, a good poet but a bad scientist."[14]

Ardent as he was in his progressionism, Huxley—working in the first half of this century—was hardly unique in this. Indeed, until late in life, Huxley drew heavily on the then-accepted paleontological picture of life's progress, of one of ever-increasing specialization leading to extinction, save for a few forms that stayed fairly general and thus paved the way from an adaptive breakthrough upward—ultimately to *Homo sapiens*. By mid-century, however, this view had come under heavy attack (with Simpson playing a major role), as it was pointed out that "special" and "general" are relative terms applied retroactively. In their way, the early mammals were highly specialized reptiles. Huxley had therefore to revise his thinking in respects, taking account of current opinion (variation in populations became much more important).

But Huxley never swerved from the belief that the distinctively human strength lies in its flexibility—the human specialization is to be generalists. With this, Huxley associated the belief that evolution outside the human realm is essentially ended. We humans alone continue to have the possibility and hope of further change—further progressive change—because we have transcended our biology, and our evolution in the psycho-social realm is no longer one of natural selection working on Mendelian genes:

> It seemed to me that, natural selection, being unable to operate only on the basis of immediate biological utility, we must accept that every tendency towards improvement of the physical machinery is bound sooner or later to reach a pitch of perfection beyond which natural selection alone cannot push it. This seems to me to apply both to "one-sided" improvements, such as those of most so-called specialists, and to general improvements in what I have called biological machinery, such as sense organs, capacity for temperature regulation, etc. The one exception to the stoppage of progressive change, namely, the human stock, no longer operates primarily by natural selection, but by the primarily psychological machinery of cumulative experience.[15]

Along with this belief, Huxley was unswerving in his commitment to the thesis that progress is an objective fact, something to be read from the biological record rather than to be read into it. The world, the organic world especially, is genuinely progressive, really and truly:

> Evolution, from cosmic star-dust to human society, is a comprehensive and continuous process. It transforms the world-stuff, if I may use a term which includes the potentialities of mind as well as those of matter. It is creative, in the sense that during the process new and more complex levels of organization are progressively attained, and new possibilities are thus opened up to the universal world-stuff.[16]

This is not just a question of humanly imposed values, or rather it does not matter if it is a question of such values: "It is immaterial whether the human mind comes to have these values *because* they make for progress in evolution, or whether things which make for evolutionary progress become significant *because* they happen to be considered as valuable to human mind, for both are in their degree true."[17]

EVOLUTIONARY HUMANISM

Since we are now into the realm of values, what of ethics? Huxley, who thought of himself as an "evolutionary humanist" and as explicitly providing a secular religion, wrote on the subject many, many times, although—drawing the convenient philosophical distinction between normative or substantive ethics ("What should I do?") and foundational or metaethics ("Why should I do that which I should do?")—for most of the time it was the former that was his chief concern. And here, naturally enough, Huxley's interests and targets changed somewhat over the years—something that he himself would have predicted and expected. Huxley always emphasized that morality is relative, meaning not that one can do as one pleases but that particular moral norms and directives evolve, especially as circumstances (like technological circumstances) change. (He even went so far as to argue that there was nothing wrong with slavery in the old days, because it was needed to keep societies running smoothly.)

However, beneath the changes, and for all that there are some particular quirks or special interests—I sense that, perhaps as a result of his years in Houston, Huxley always had a somewhat condescending attitude toward Negroes and their abilities[18]—a consistent pattern or theme emerges. While Huxley was not uninterested in life at the personal level, it was the general domain that really excited him:

> All claims that the State has an intrinsically higher value than the individual are false. They turn out, on closer scrutiny, to be rationalizations or myths aimed at securing greater power or privilege for a limited group which controls the machinery of the State.

> On the other hand the individual is meaningless in isolation, and the possibilities of development and self-realization open to him are conditioned and limited by the nature of the social organization. The indi-

vidual thus has duties and responsibilities as well as rights and privileges, or if you prefer it, finds certain outlets and satisfactions (such as devotion to a cause, or participation in a joint enterprise) only in relation to the type of society in which he lives.[19]

The key moral principle seems to have been for the need of planning in running the state and, above all, the application of *scientific* principles and results in such planning and its implementation. You simply cannot (or should not) leave things to chance or intuition—the implication being that this is precisely where your average politician does leave things—but should bring the trained scientific mind to bear on life's problems.

Again and again Huxley returned to this theme. For instance, in a book that he wrote in the interwar years, *If I Were Dictator*,[20] he stressed the need for science in the running of an efficient state and that such science would need to be of the social variety as well as physico-chemical and biological. During the Second World War, he wrote a highly laudatory essay on the Tennessee Valley Authority, that marvel of the Rooseveltian New Deal, whereby the federal government built and ran a massive system of river damming and irrigation in what had hitherto been one of the more desolate parts of the US.[21] Then, after the war it was Huxley who insisted on science being added to UNESCO, and he wrote a vigorous polemic arguing that the organization had to be run on evolutionary lines—lines demanding lots of science.[22] So vigorous was his polemic, indeed, that he upset his masters and he was refused a full four-year term as director general.

Now let us turn to the question of justification, that of metaethics. With justice one might argue that much that Huxley believed and preached came straight from his personal experiences and prejudices. An enthusiasm for eugenics, for instance, was part of the general credo of intellectuals in the first parts of this century. Likewise with birth control, another key plank in the Huxley scheme of things. Huxley had known and socialized with leaders of the movement in the early years. And obviously the enthusiasm for science stemmed straight from the raison d'être of Julian Huxley. Here, the fact that he was an evolutionist was contingent, if not irrelevant. In principle, he could have believed in the literal truth of Genesis, so long as he was enough of a physicist or some other scientist to want to advocate the all-importance of science in running the state.

However, Huxley certainly thought he was putting his substantive ethics on an evolutionary metaethical base (though he does not use this language) and gave arguments to this effect. Most explicit was the discussion that came in 1943 (one year after *Modern Synthesis*) in a Romanes lecture—"Evolutionary ethics"—at Oxford, something he published commercially a few years later (bound together with the Romanes lecture, also on ethics, given by his grandfather some fifty years before). At this point, Huxley was rather enthused by Freudian theory, so one has to wade through introductory filler on the psychoanalytic development of the moral sense before one gets to the biological meat. But it is there, unambiguously.

And it is there that one sees how terribly crucial was biological progressionism for this key aspect of Huxley's "Religion without Revelation."[23] Simply, Huxley argued that since evolution is progressive and since progress means that value is ever increasing, we have a moral obligation to cherish and to promote the evolutionary scheme of things:

> When we look at evolution as a whole, we find, among the many directions which it has taken, one which is characterized by introducing the evolving world-stuff to progressively higher levels of organization and so to new possibilities of being, action, and experience. This direction has culminated in the attainment of a state where the world-stuff (now moulded into human shape) finds that it experiences some of the new possibilities as having value in or for themselves; and further that among these it assigns higher and lower degrees of value, the higher values being those which are more intrinsically or more permanently satisfying, or involve a greater degree of perfection.

> The teleologically-minded would say that this trend embodies evolution's purpose. I do not feel that we should use the word "purpose" save where we know that a conscious aim is involved; but we can say that this is the *most desirable* direction of evolution, and accordingly that our ethical standards must fit into its dynamic framework. In other words, it is ethically right to aim at whatever will promote the increasingly full realization of increasingly higher values.[24]

We see now why it was so crucial for Huxley to maintain that evolution, except in our species, is essentially finished. It was no part of his brief to argue that we must promote the evolution of (say) the warthog, except inasmuch as such animals or plants benefit us humans. It was also crucial to maintain that in humans the mode of evolution has changed. Huxley was not about to get snared into arguing for some kind of laissez-faire philosophy, either at the biological or social levels. Indeed, he opposed it ardently. He was for planning, and if not a socialist (certainly not a social democrat) he was for the use of the state to effect major schemes and projects. Remember the TVA.

Yet equally, the characterization of progress in terms of control and independence was a vital part of Huxley's argument, the positive part. If the need is to promote progress and such progress consists in maximizing control and independence, what better way to do it than through state-organized schemes for the implementation of the fruits of science and technology? Although Huxley may well have agreed that it was the spur of circumstances and acquaintances that led him initially to stress the need for such things as artificial birth control and scientific agriculture and the like, he would have argued that his justification for such norms was entirely logical—or rather, entirely evolutionary. Without such things, real progress in the human realm becomes impossible.

The argument therefore is that the evolutionary process itself promotes value, and therefore our ethical duty is to work with and within this process to see that it is realized as fully as possible:

> From the evolutionary point of view, the destiny of man may be summed up very simply: it is to realize the maximum progress in the minimum time. That is why the philosophy of Unesco must have an evolutionary background, and why the concept of progress cannot but occupy a central position in that philosophy.[25]

"New bottles for new wine" Huxley labeled one of his collections of essays.[26] In this case, however, the wine is old, however fancy and new looking the container. One can readily link Huxley's views with a long-standing—the most long-standing—tradition in evolutionary ethics, as his critics were very quick to point out. Grant for the moment the distinctive features of Huxley's position, for instance, about human evolution being uniquely that which is still progressing (although more on that shortly), the underlying pattern (Greene is right in this) goes back to Herbert Spencer. He, too, was an ardent evolutionary progressivist—"monad to man"—believing that the philosophy is at one with progressionism in the sociocultural human world. He, too, wanted to replace conventional religion with some sort of evolutionary substitute—the "Synthetic Philosophy." And he, too, saw normative directives in evolution's methods, arguing that we humans have a moral obligation to promote the ends of evolution, and that the reasons for this lie in the process itself. It is precisely because evolution is progressive that we ought to cherish it, for here and only here is the way to promote and preserve value.

For all that Bergson was the primary nonbiological influence, I doubt Huxley would have been particularly perturbed by my locating him in the Spencerian tradition (apart from the fact that Bergson acknowledged a debt to Spencer). There was ever a Spencerian flavor to Huxley's work—"A state of equilibrium may for a time exist, but every balanced organism is as it were pressing against every other, and a change in one means a rearrangement of them all"[27]—and when it came to progress, Huxley was explicit in his praise that in seeing uniformity ("homogeneity") give way to complexity ("heterogeneity"), Spencer had hit on an idea containing "a great deal of truth."[28]

Since traditions move forward as well as backward, let me point out also that Julian Huxley was not the end point to the line of those who argue in neo-Spencerian manner. To choose but one, in recent years the Harvard entomologist and sociobiologist Edward O. Wilson has been endorsing and promoting an evolutionary ethical picture very much in line with that of Huxley.[29] Wilson's Categorical Imperative beseeches us to promote biodiversity—an issue dear to the heart of one whose speciality is tropical ants and who is keenly aware of the ongoing destruction of the Brazilian rain forests—but his justification is the upwardly progressive nature of the evolutionary process. Significantly, on the wall of Wilson's laboratory hangs a picture of Herbert Spencer and in his recent autobiography Wilson speaks of Huxley with much respect.[30]

CRITICAL RESPONSE

What can one say about Huxley's moral philosophy? Let me now step out of my role as disinterested historian and into my role as partisan philosopher. One thing is already obvious, if further confirmation were necessary. The actual normative directives endorsed and promoted by evolutionary ethicists are very far from the evil philosophies of legend. I am not sure that I particularly fancy having Julian Huxley as my dictator, but far rather him than some of the real ones. It is true that today some of his prescriptions have an uneasy ring—eugenics, for instance, although my reading on that particular issue suggests that Huxley was more in favor of what today we call "genetic counseling" (that is, elimination of bad genes) than of wholesale biological redesign. Likewise, nowadays, major state-run projects have fallen out of favor; although in this case there are many (myself included) who think that there is still much to be said in their favor. We also think that a major reason why such projects are out of favor is that they have served their function and that now with increased prosperity they are no longer needed. This, of course, speaks only to Western society. Elsewhere in the world Huxley-like projects are both needed and promoted.

Grant that a case can be made for Huxley-style normative prescriptions and that, even more strongly, can a case be made for the more recent prescriptions of those in the same tradition. Biodiversity and the preservation of the rain forests is close to being a motherhood issue. But what of the question of justification? What about the argument that Huxley gives to support his normative ethics? What about the appeal to the progressive nature of the evolutionary process? It is here that Moore faulted Spencer, and likewise it is here that he would have faulted Huxley, accusing them both of having committed the naturalistic fallacy.

In the case of Huxley, the charge would be that control and independence—together with complexity, which is presumably a condition of these—are natural properties or phenomena, things of this world as it were. However, the directives that one ought to promote control-abilities and independence (through science and planning) are nonnatural, in the sense that they do not derive from the physical realm—although obviously we may want to apply them to or in such a realm. Hence, the former simply cannot be the foundational base of the latter. Or, if we were to phrase things in the manner of David Hume—who argued that there is a logical gap between claims about matters of fact ("is" statements) and claims about morality ("ought" statements)—the charge would be that Huxley goes illicitly from the way that things are (in the evolutionary process) to the way that things should be (in the future). He goes from "control and independence exist," to "control and independence should be maximized."

Although I am not aware that Huxley did address this criticism directly (very much unlike Simpson, as we shall learn shortly), one ready response would have been to agree that there certainly is (or seems to be) a significant difference between claims about fact and claims about morality, but that this in itself is no barrier to the former being the foundation of the latter. After all, in science we are very used to "reductions": derivations or justifications of one thing in terms of another very different. Is the is/ought distinction any greater than (say) the micro/macro division to be found in the theory of gases? One goes there from the temperature-free molecules buzzing in empty space to a gas occupying all of the space in a container—a gas that can go from very cold to very hot and that can possibly have a tremendous pressure into the bargain. Yet the one is supposed to be the foundation of the other, and there are mathematical deductions to prove it!

Perhaps Huxley does make some sort of response along these lines, for it is worth remembering that he stresses repeatedly that he sees the progressiveness of evolution as an entirely natural, objective phenomenon. Hence, for him, the value-laden nature of independence and control is not something read into nature, but something read from it. Thus the values are given to us by the natural world, and so at least logically there is no reason why I should not go from an "is" statement to an "ought" statement. I will return to this point in a moment; but, in any case, first let me agree with Huxley if—in whatever way—he does simply refuse to bow down before the above critical argument. Why should it be that, in going from natural to unnatural, he is thereby committing a "fallacy"? What right have his critics to label his argument fallacious—without further argument on their part, that is? After all, we do value the things that make us distinctively human, so why should we not promote them?

To his credit, a point often missed by those who quote him favorably, to back his contention that Spencer's ethic was inadequate, Moore did offer further detailed critique. And it is worth noting that, although Huxley's arguments found no greater support among professional philosophers, they, too, felt it is necessary to take time to refute him explicitly, also. In particular, the Cambridge moral philosopher C. D.

Broad (former student of Moore) subjected Huxley's thinking on the subject to withering critical review. About the move from "is" to "ought" he was strong in his objection. Raising the key question of "whether knowledge of the facts of evolution has any bearing on the question of what is intrinsically good or bad," Broad wrote:

> It is plain that Prof. Huxley thinks that it has an important bearing on this question, but I find it extremely hard to see why he does so. Perhaps I can best bring out the difficulty that I feel in the following way. Take the things which Prof. Huxley considers to be intrinsically good, and imagine him to be confronted with an opponent who doubted or denied of any of them that it was intrinsically good. How precisely would he refute his opponent and support his own opinion by appealing to the facts and laws of evolution? Unless the notion of value is surreptitiously imported into the definition of 'evolution,' knowledge of the facts and laws of evolution is simply knowledge of the *de facto* nature and order of sequence of successive phases in various lines of development.[31]

Broad continued:

> If then, Prof. Huxley is to support his own views about the intrinsic value of so-and-so and to refute those of an opponent by appealing to the facts and laws of evolution, there must be a suppressed premise in the argument. . . . He must use *some* 'mixed' premise, connecting certain *purely factual* characteristics, which are all that a study of evolution can possibly reveal to us, with the *value-characteristics* of intrinsic goodness and badness.[32]

Broad concluded:

> Now, whatever may be the evidence for such a mixed premise, it is quite plain that it must be something different from the evidence for the facts and laws of evolution. For the premise required asserts a connection between certain of those facts and laws and something else, viz., intrinsic value or disvalue, which forms no part of their subject-matter.[33]

Presumably Huxley's response at this point would be a reiteration of the point made above, that there is no reason in principle why one should not sometimes—this being one of those sometimes—go from claims of fact to claims of value. But I think Broad is trying to dig more deeply here. If Huxley raises the reduction analogy—one can legitimately go from micro to macro in gas theory—the reply is that one can certainly do so *if* one can give good reasons why this might be allowed. In gas theory, good reasons are given. In Huxley's case, no such reasons have been given, and there are at least two major counter-reasons to think there are no such reasons.

On the one hand, although Broad (as a nonbiologist) does not really argue this, there is the question of Huxley's picture of the evolutionary process. This is crucial, or so it should be if one is promoting an ethics based on and informed by evolution. But, whether one takes Huxley's original position or the later revision, there are serious questions. It is highly dubious whether the process of evolution has come to an end save in the human line, and that consequently the only evolution is a nonselective sociocultural process. More than this, there is no reason to think that evolution promotes control and independence in the manner supposed by Huxley, particularly in the manner that supposes that humans have some special status in this respect—a status that allows (requires) us to speak of progress as a matter of objective fact. Humans have lost the ability to synthesize vitamin C. Does this mean that we are any less worthy than formerly, unless we want to make long sea voyages? Females have lost the ability to reproduce without the aid of males. Does this make them less valuable on the scale of being? Or conversely, if males really do control females, any more worthy? Sometimes independence and control are good things; sometimes not. It all depends. Whales can dive deeply and live for long under water; birds can fly; camels can exist in the desert without water. We can do none of these things. Although we have less control and independence, are we thereby lesser beings?

On the other hand, the point that Broad does stress, suppose someone denies that control and independence are so very all-important and valuable. There is nothing in evolution, certainly nothing in Darwinian evolution as such, to convince you otherwise. You may of course challenge their values, if for instance they prize weapons of attack for use in life's struggles, but you are not doing this on the basis of evolution. Rather, these are values that you import from elsewhere. The fact of the matter is that Huxley simply reads in what he wants to read out. Huxley wants control and independence of a peculiarly human variety to count, so he simply uses them as a lens through which to view evolution. The inescapable truth is that there is a crucial gap between premises and conclusion in Huxley's argument for the foundations of ethics. State planning may be all very well—in my opinion, it is all very well—but it has been assumed and not proven. It has certainly not been proven from evolutionary premises, even if one agrees one can get it from control and independence.

So much then for Huxley. Almost necessarily in a discussion such as this, one comes through sounding extremely negative, so let me conclude by going back to the beginning point, a shared conviction with both Julian and his grandfather that it really must matter that we are evolved beings and not just the product of a Good God on the Sixth Day, molded in His image. Julian Huxley and his tradition seem not to be on the right track. The question is whether, as people like Broad believed, that is the end of evolutionary ethics. Simpson thought not, so let us turn now to look at his thinking.

GEORGE GAYLORD SIMPSON (1902–1984)

Simpson was a genius. The best paleontologist since Cuvier, his supervisor called him. Educated at Colorado and Yale, Simpson spent much of his working life at the American Museum of Natural History in New York, moving later to Harvard and then retiring to New Mexico. He worked extensively on the mammals and more generally, having a command of mathematics unusual among evolutionists, it was he who was primarily responsible for bringing the fruits of the Darwin-Mendel synthesis to bear on the fossil record. His masterwork *Tempo and Mode in Evolution* (revised and retitled in 1953 as *Major Features of Evolution*) appeared in 1944, two years after Huxley's *Modern Synthesis*, although it was completed at about the same time. A shy and difficult man, Simpson had few close friends, but in later life Huxley—who had been friendly to the young man when he was a postdoctoral fellow in London after Yale—became very intimate, discussing in detail questions of evolution and ethics.[34]

The place to start is with a point made earlier, that a major mission in Simpson's life was to shatter the picture of progressive change absorbed and endorsed by Huxley. Simpson denied that the past was anything at all like that. It was far too much given to branching for any Huxley-like scenario; the talk of the specialized being always dead ends was simply not true; and natural selection is too opportunistic for anyone to suppose that independence and control have the privileged positions presumed by Huxley. Simpson could never have written: "One somewhat curious fact emerges from a survey of biological progress as culminating for the evolutionary moment in the dominance of *Homo sapiens*. It could apparently have pursued no other general course than that which it has historically followed."[35] For the American, evolution is quite undirected, driven only by the immediately adaptively advantageous. And it is simply false to think that evolution has stopped or that it is now confined to the human realm. There are all sorts of possibilities for further change.

> One cannot doubt that a *particular channel* of specialization may reach its limit (the horse cannot have less than one toe), but then progress (or change, at least) shifts to some other channel through, as a rule, some other group of animals. Even if, as you suggest, all systems except the nervous system were near the limit of biological efficiency by the end of the Miocene, the nervous system did, in our time, go on from there, which was hardly predictable at the time. I cannot feel any confidence in prediction that the nervous system is now at its limit, or even that quite other possible channels of progress, in the nervous system or in other systems, are really non-existent.[36]

However, this much said—and it is to say a lot—in Simpson's writings there are certainly strong echoes of Huxley's thought, which (Simpson candidly admitted) acted as a major stimulus to his own thinking. Most particularly, Simpson was ardently and unambiguously a progressionist, and he thought it simply silly not to agree that humans are in basic respects at the top of the mountain (or "pinnacle" as he and everyone else tended to call it, following Sewall Wright). He believed that there are ecological niches existing objectively, waiting to be occupied, and that organisms make adaptive breakthroughs in moving to take them over. And he certainly believed that there was some sort of hierarchy, of a not entirely unfamiliar kind.

Yet—and here we run straight into one of the most crucial aspects of Simpson's thought—he was keen to stress that although the denial of progress was silly, the ascription of progress was not something simply read from nature in an objective manner (as Huxley thought). Rather, it is something that we must read in, and since we must avoid the gross anthropomorphism of simply reading in criteria that are by definition human-like, this means that taken generally there can be no one mark of progress that is especially privileged over all others.

For this reason, in his most extended discussion of the topic (in a popular book, *The Meaning of Evolution*, published in 1949), Simpson ran through a large number of proposed criteria of biological improvement or worth: expansion of life, dominance, specialization, potential for future development, independence from the environment, control of the environment, complexity (Simpson was not very keen on this old favorite, believing that it is probably false that complexity increases, and is an untestable concept anyway), general energy level, pre- and postnatal care, sophisticated nervous system, individualization, and more.

Humans certainly do not come out top on all of these. For instance, Simpson was sufficient of a traditionalist to think that we humans are not very specialized. However, overall, we tend to score well, the very best in many cases—like dominance and pre- and postnatal care and nervous system and individualization. And this general consilience seems to have been enough to convince Simpson that progress, with humans at the top, was more than just a whim or conceit. Progress may not have had the objectivity that Huxley supposed, but it was more than just one man's yearning.

Move on now to ethics and again we see a reaction against the kind of thinking that Huxley represented, although even here again it was a sympathetic reaction. Like Huxley, Simpson was absolutely and completely committed to the view that ethics is natural, in the sense of being produced by evolution. In fact, Simpson was rather more detailed than Huxley on the way in which biology through the medium of selection could produce something like an ethical sense, pointing out that success in the struggle for existence does not necessarily mean all-out warfare but can demand sympathetic alliance with one's fellows. (It is true that Simpson was more inclined to a group perspective on the workings of evolution than most evolutionists would find acceptable today.) Ethics is natural also in having no justification or sanction outside of evolution.

What has evolved is what you get. Simpson, who came from a fundamentalist Presbyterian family, was by nature (like both Huxleys) always an intensely religious man. But his faith in an existent deity was nonexistent (in middle life he worshipped with the Unitarians), and he certainly thought there could be no divine or similar support for moral belief.

At the level of normative ethics—"What should I do?"—whereas Huxley was in favor of large-scale public works and other state-funded projects, Simpson looked much more to the individual level. There were two major directives. First, there was the need to improve and promote knowledge—knowledge in itself, as a good:

> The most essential material factor in the new evolution seems to be just this: knowledge, together, necessarily, with its spread and inheritance. As a first proposition of evolutionary ethics derived from specifically human evolution, it is submitted that promotion of knowledge is essentially both the acquisition of new truths or of closer approximations to truth (metaphorically the mutations of the new evolution) and also its spread by communication to others and by their acceptance and learning of it (metaphorically its heredity). This ethic of knowledge is not complete and independent. In itself knowledge is necessarily good, but it is effective only to the degree that it does spread in a population, and its results may then be turned by human choice and responsible action for either good or evil.[37]

Then secondly we have personal responsibility, which leads to integrity and dignity:

> Beyond its relationship to the ethic of knowledge, the fact of responsibility has still broader ethical bearings. The responsibility is basically personal and becomes social only as it is extended in society among the individuals composing the social unit. It is correlated with another human evolutionary characteristic, that of high individualization. From this relationship arises the ethical judgment that it is good, right, and moral to recognize the integrity and dignity of the individual and to promote the realization or fulfilment of individual capacities. It is bad, wrong, and immoral to fail in such recognition or to impede such fulfilment. This ethic applies first of all to the individual himself and to the integration and development of his own personality. It extends farther to his social group and to all mankind. Negatively, it is wrong to

develop one individual at the expense of any other. Positively, it is right to develop all in the greatest degree possible to each within the group as a whole. Individuals vary greatly in other capacities, but integrity and dignity are capable of equal development in all.[38]

Why the differences with Huxley? To be honest, I doubt that they had much to do with biology, despite the supposed derivation of these ideas from evolution. In Huxley's case, I have suggested that the personal factor was significant, his acquaintances and the like, although one should not forget his upbringing—the worth of science (and of a scientific elite) in the Huxley household—and the challenges (and successes) of society, particularly society in the 1930s, when Keynsian economics seemed to provide the way forward from widespread depression and poverty. In Simpson's case, the valuing of knowledge was equally personal. In the 1930s, seeking for an alterative to conventional religious faith, he had found it in the search for knowledge—something that, incidentally, helped to salvage his conscience as he regularly dumped his four daughters (Simpson was a single father) on his parents, as he would go off on long-scale fossil hunting trips.

The valuing of responsibility and dignity and so forth was equally a function of the times and society within which Simpson lived. We are talking now of the years when the cold war was settling right into its long winter, when Soviet science was suffering under influential charlatans like Lysenko, and when issues of dictatorship and totalitarianism were all too fresh in people's memories and present in much of the world of the day. From dignity and responsibility, Simpson launched straight into a condemnation of the oppressive regimes then flourishing, and he juxtaposed this with a cherishing—if not an uncritical cherishing—of the society within which he found himself: "Democracy is wrong in many of its current aspects and under some current definitions, but democracy is the only political ideology which can be made to embrace an ethically good society by the standards of ethics here maintained."[39]

SIMPSON'S METAETHICS

What of justification? Wherein lay the foundations of normative ethics for one such as Simpson? It is not easy to give a direct answer to this question, a fact that reflects that the usually very clear Simpson was himself wrestling with his response. He knew, on the one hand, that the answer had to come from evolution. That was the whole point of naturalism. He knew on the other hand—and, unlike Huxley, Simpson took this very seriously—that one simply cannot go blithely from statements about facts, "is" statements, to statements about obligations, "ought" statements. The question is, how do you reconcile these two demands or constraints?

At times, Simpson sounded almost like an existentialist in his thinking. Could it not be that the very fact of responsibility, of the need to make a choice in one's own right, is in some way self-validating? Ethics is certainly relative to the material and social situation in which one finds oneself. Could it not be that they are relative also to the individual—the evolved individual, that is?

> It should, finally, again be emphasized that these ethical standards are relative, not absolute. They are relative to man as he now exists on the earth. They are based on man's place in nature, his evolution, and the evolution of life, but they do not arise automatically from these facts or stand as an inevitable and eternal guide for human—or any other—existence. Part of their basis is man's power of choice and they, too, are subject to choice, to selection or rejection in accordance with their own principles. They are also subject to future change as man evolves; after all, if mankind does pursue the ethic of knowledge it should be able progressively to improve and refine any ethical system based on knowledge.

> There is no ethical absolute that does not arise from error and illusion. These relativistic ethics have, at least, the merit of being honestly derived from what seems to be demonstrably true and clear.[40]

The trouble with such a position as this is that, ultimately, there is no arbiter between right and wrong. One is left with the subjectivism of the individual conscience. Yet, although Simpson had trouble with individual relationships, he had a burning sense of the imperatives of morality, imperatives that drove him to action. As Huxley worked for and defined UNESCO, so Simpson during the Second World War had engaged in difficult and dangerous military service—at an age when he could have excused himself—and later in life he was outspoken in his moral disgust at America's involvement in Vietnam. There was not much of the relativistic here.

Simpson clearly went on worrying and, in the 1960s, returned to the issue of foundations. It is worth quoting him at length.

> The evolutionary process in itself is nonethical—there simply is no point in considering whether it is good, bad, a mixture of the two, or neither. As would be expected, some mammals other than man do seem to have slight rudiments of a moral sense, but on the whole it is nonsensical to speak of ethics in connection with any animal other than man, and still more so in connection with the plant kingdom—which after all is just as much a part of evolutionary history and trends as the animal kingdom. No, *man* is the only ethical organism in the full meanings of the words, and there are no relevant ethics except human ones. But man, with his ethical sense, was produced by evolution as the result of one evolutionary sequence. Human evolution continues and man is subject to all evolutionary principles, not simply as generalists but as they apply in his specific and extremely special circumstances. A rational, naturalistic system of ethics cannot be independent of evolution, but neither can it be derived from evolution as a general or merely abstract principle.[41]

He added:

> Biological trends are almost entirely directed by natural selection. Social trends operate within the framework and limitations of the biological factors and represent a remarkably flexible form of biological adaptation. . . . The point pertinent here . . . is that neither strictly organic nor social evolution necessarily leads to improvement in any sense of the word acceptable for the human situation. In fact either, in their ways that *in themselves* are blindly amoral, may have results extremely undesirable for mankind. But here is the saving grace: both organic and social evolution are now to some degree, even though a limited degree, within our own control. We, alone among all organisms, are aware of our own sociobiological evolution, can judge what seems to us desirable or undesirable in that evolution, and can deliberately work for the desirable.

> If ethics could be reduced to ethicizing, authority acceptance, and nothing else, then such discussions as this would be a mockery. Unless

we reject naturalism entirely, that would leave us firmly impaled on the naturalistic fallacy that what is ought to be. There is really no point in discussing ethics, indeed one might say that the concept of ethics is meaningless, unless the following conditions exist: (*a*) There are alternative modes of action; (*b*) man is capable of judging the alternatives in ethical terms; and (*c*) he is free to choose what he judges to be ethically good. Beyond that, it bears repeating that the evolutionary functioning of ethics depends on man's capacity, unique at least in degree, of predicting the results of his actions. A system of naturalistic ethics then demands acceptance of individual responsibility for those results, and this in fact is the basis for the origin and function of the moral sense.[42]

Ultimately, it all comes down to evolution, and for this reason sympathetic critics of Simpson have concluded that, despite his intentions, ultimately he, too, goes the way of Huxley, justifying morality in the name of evolution and hence, at the end, committing the naturalistic fallacy. Yet, I have to say that I disagree—he speaks even here too strongly against the move of justifying morality by reference to the facts of evolution—although I would agree that Simpson gives no adequate analysis, and indeed at the end of the passage just quoted seems to slide back again into an unsatisfactory subjectivity. The murderer might accept responsibility, but that is no excuse.

Is there a way out of this dilemma? Speaking now as a philosopher rather than a historian, I believe there is, although whether Simpson would have been happy with it is another matter. As a number of thinkers have argued in recent years, one might claim that inasmuch as one believes that the moral sense is a product of evolution, this suggests that ultimately there is *no* foundation for ethics considered at the normative level! This does not mean that ethics does not exist, or that one can simply do what one pleases—at least, one cannot simply do what one pleases inasmuch as one is an evolved human being. What it means is that at the foundational, metaethical level, there is no ultimate justification, whether this be evolution or the Will of God or whatever else. (I have defended this position in my *Taking Darwin Seriously: A Naturalistic Approach to Philosophy* and *Evolutionary Naturalism: Selected Essays*.)

According to the "ethical skeptic" therefore, ethics—normative ethics—has evolved to make us good cooperators because, given the kinds of beings we humans are, cooperation is a good adaptive strategy in the struggle for existence. But there is nothing beyond this, and certainly no solid ground of proof. We have a moral sense, because it is adaptively advantageous to have it, but ultimately—like the secondary qualities that appear so vividly to us—there is nothing that it is sensing! Hence, a causal explanation is all that you can give, and once you have given it, you see that the call for a reasoned justification is impossible (although, thankfully, not necessary) to satisfy:

> The [evolutionist] may well agree . . . that value judgments are properly defended in terms of other value judgments until we reach some that are fundamental. All of this, in a sense, is the giving of *reasons*. However, suppose we seriously raise the question of why these fundamental judgments are regarded as fundamental. There may be only a *causal* explanation for this! We reject simplistic utilitarianism because it entails consequences that are morally *counterintuitive*, or we embrace a Rawlsian theory of justice because it systematizes (places in 'reflective equilibrium') our *pretheoretical convictions*. But what is the status of those intuitions or convictions? Perhaps there is nothing more to be said for them than that they involve deep *preferences* (or patterns of preference) built into our biological nature. If this is so, then at a very fundamental point the reasons/causes (and the belief we ought/really ought) distinction breaks down, or the one transforms into the other.[43]

It is usually added by people who argue this way that ethics works because we are part of the system. With a shared evolution, we humans have a shared insight—or rather, sense of insight—into the norms of right and wrong. Some people may disagree with these norms, but as children make mistakes in arithmetic, the disagreement is a function of inadequate training and does not point to an irresolvable subjectivity. Or the disagreement is biologically based. Just as some are unable to tell red from green, but that is their misfortune and not to be laid at the nonexistence of a real difference between red and green, so some are innately morally handicapped. What is crucial is that, although under this picture, morality ultimately has no objective existence (in the sense of an objective referent—it is objective in that it certainly exists), our biology makes us think that it does! We believe we are talking about something real, because if we did not, morality would collapse into a morass of subjective wishes and desires and directives, and it would entirely lose its force.

I like to think that Simpson would respond favorably to the spirit of this solution. One is a firm evolutionary naturalist. At the same time, one avoids the naturalistic fallacy because one is not justifying anything, including ethics, in terms of anything else, including evolution! And although personal responsibility and choice are important—the whole point of this view of ethics is that one does have the freedom to choose and is not locked into one course of action—and, although it is certainly agreed that norms will change as circumstances change, ethics is more than just personal feelings. There are standards which are society-, if not human-wide.

However, reverting back to the historian, I cannot pretend that it is a solution that Simpson himself achieved or endorsed, certainly not in its entirety. There is nothing in Simpson, for instance, about the lack of foundations for normative morality. Nor is there anything about having a sense that there are foundations, since it is adaptive to have such a sense. Indeed, to be candid, Simpson might well have pulled back from such a solution believing that, ultimately, one is downgrading ethics (normative ethics, that is) by claiming it to be (what I myself have elsewhere characterized as) an "illusion of the genes."[44] In defense of the solution, one can explain at length that the illusion comes, not in the nonexistence of ethics as such, but in the supposed referent. But this might not have convinced Simpson, or at least only in mind and not in heart.

Before I leave the discussion, however, let me refer back to Greene's claim, now to the part about Simpson being the latter-day representative of Thomas Henry Huxley. There is certainly good reason for agreeing with this, as well as with the implication that Huxley grandfather and Simpson represent a rival tradition to that of Spencer and Huxley grandson. Although Simpson did not regard the products of evolution to be quite as vile as did T. H. Huxley, they agreed that one must strive for a naturalistic approach to ethics, they agreed that this approach must be evolutionary and nothing else, and they agreed that one cannot justify ethics in the name of evolution—Huxley was eloquent of the degree to which evolution is not necessarily progressive in any conventional sense.

Given Simpson's fuzziness, can we usefully ask about T. H. Huxley's metaethical philosophy? Where did he think one should locate the foundations of substantive ethics? Unfortunately, like Simpson, there is (from another usually crystal clear thinker) less than total clarity. But interesting and informative is a little book that Huxley wrote on David Hume—interesting because, although no evolutionist, Hume's metaethics is a clear forerunner of the skeptical solution I have sketched just above, and informative because Huxley seems to respond warmly to it. Toward the end of the discussion of Hume's moral theory, what starts as paraphrase seems to conclude as whole-hearted agreement:

> In whichever way we look at the matter, morality is based on feeling, not on reason; though reason alone is competent to trace out the effects of our actions and thereby dictate conduct. Justice is founded on the love of one's neighbour; and goodness is a kind of beauty. The moral law, like the laws of physical nature, rests in the long run upon instinctive intuitions, and is neither more nor less "innate" and "necessary" than they are. Some people cannot by any means be got to understand the first book of Euclid; but the truths of mathematics are no less necessary and binding on the great mass of mankind. Some there are who cannot feel the difference between the "Sonata Appassionata" and "Cherry Ripe;" or between a grave-stone-cutter's cherub and the Apollo Belvidere; but the canons of art are none the less acknowledged. While some there may be, who, devoid of sympathy, are incapable of a sense of duty; but neither does their existence affect the foundations of morality. Such pathological deviations from true manhood are merely the halt, the lame, and the blind of the world of consciousness; and the anatomist of the mind leaves them aside, as the anatomist of the body would ignore abnormal specimens.[45]

Even if Huxley is speaking entirely for himself—and I am inclined to think that he is, because a few years after he wrote of "a keen innate sense of moral beauty,"[46] which is in itself the means and the end of morality—this is hardly a full-blooded exposition and endorsement of ethical skepticism. But it is in the spirit of it, not just in being naturalistic but in so firmly putting morality in the domain of feelings—something that the evolutionist like Simpson would relate back to natural selection and adaptation. So, without pretending that a definitive case has been made, it seems not too unfair to suggest that just as the tradition of Spencer and Julian Huxley has been taken up today by people like Wilson, so the tradition of T. H. Huxley and Simpson has been taken up today by those evolutionists who think their theory points to ethical skepticism.

CONCLUSION

My conclusions follow quickly. As a historian I find that Julian Huxley and George Gaylord Simpson were united in believing that morality must be natural and that the science of key significance was evolution. They represent, however, two different traditions, going back (at least) to Herbert Spencer and Thomas Henry Huxley. Julian Huxley wanted to justify normative ethics by reference to the fact and processes of evolution, and to do this he had to argue that progress is an objective fact of nature. He was therefore prepared to barge right through the naturalistic fallacy. Simpson wanted to explain normative ethics in terms of the fact and processes of evolution and in doing this, although a progressionist, he wanted to avoid claiming that this is an objective fact of nature. He took the naturalistic fallacy seriously, as something to be avoided, and talked much of the significance of individual responsibility as a foundation for morality.

As a philosopher I applaud the spirits and intentions of both men. I take this applause as showing an agreement stronger than any differences that we may have. I am happy to be set off with the evolutionists against the tradition of twentieth-century analytic philosophy, typified by G. E. Moore. However, I do agree that Julian Huxley's position is inadequate, and I would argue with the critics (in this paper with C. D. Broad but ultimately with David Hume) that there is a key difference between statements of fact and statements of morality and that Huxley and his tradition fail to bridge it. Progress is not an objective fact of nature and cannot therefore be used to justify a normative ethic.

I find more encouraging Simpson's thinking about morality, and that is true more generally of his science and the interpretations he would put into it. I do not think that Simpson himself gives a full and satisfying metaethics, although I would defend him against critics who claim that he commits the same fallacies as does Huxley. Whether Simpson would appreciate the directions in which I point him is another matter. This, however, must be left to the judgment of the reader.

NOTES

1. C. E. Russett, *Darwin in America: The Intellectual Response. 1865–1912* (San Francisco: Freeman, 1976); R. Bannister, *Social Darwinism: Science and Myth in Anglo-American Social Thought* (Philadelphia: Temple University Press, 1979); G. Jones, *Social Darwinism and English Thought* (Brighton, UK: Harvester, 1980); R. J. Richards, *Darwin and the Emergence of Evolutionary Theories of Mind and Behavior* (Chicago: University of Chicago Press, 1987); M. Ruse, *The Evolution-Creation Struggle* (Cambridge, MA: Harvard University Press, 2005).

2. D. Hume, *A Treatise of Human Nature* (Oxford: Oxford University Press, 1978).

3. H. Sidgwick, *The Methods of Ethics* (London: Macmillan, 1874).

4. T. H. Huxley, "Evolution and Ethics," in *Evolution and Ethics* (London: Macmillan, 1893), pp. 46–116.

5. Letter to F. Dyster, Huxley Papers, 15.106.

6. J. S. Huxley, *Memories* (London: Allen and Unwin, 1970), *Memories II* (London: Allen and Unwin, 1973); C. K. Waters and A. van Helden, eds., *Julian Huxley: Biologist and Statesman of Science* (Houston: Rice University Press, 1992).

7. J. S. Huxley, *The Courtship Habits of the Great Crested Grebe* (London: Jonathan Cape, 1914), *Problems of Relative Growth* (London: Methuen, 1932).

8. J. S. Huxley, *Evolution: The Modern Synthesis* (London: Allen and Unwin, 1942), p. 559.

9. Ibid., pp. 564–65.

10. Ibid., p. 575.

11. J. S. Huxley, *The Captive Shrew and Other Poems of a Biologist* (Oxford: Basil Blackwell, 1932), p. 53.

12. J. S. Huxley, *The Individual in the Animal Kingdom* (Cambridge: Cambridge University Press, 1912), pp. 6–7.

13. M. Ruse, *Monad to Man: The Concept of Progress in Evolutionary Biology* (Cambridge, MA: Harvard University Press, 1996), p. 612.

14. J. S. Huxley, *Evolution: The Modern Synthesis*, pp. 547–48.

15. Letter to G. G. Simpson, December 4, 1950. In Huxley Papers, Rice University.

16. T. H. Huxley and J. S. Huxley, *Evolution and Ethics 1893–1943* (London: Pilot, 1947), p. 131.

17. J. S. Huxley, *Essays of a Biologist* (London: Chatto and Windus, 1923), pp. 59–60.

18. J. S. Huxley, "The Negro Problem," *Spectator*, November 29, 1924, pp. 821–22.

19. J. S. Huxley, *What Dare I Think? The Challenge of Modern Science to Human Action and Belief* (London: Chatto, 1931), pp. 138–39.

20. J. S. Huxley, *If I Were Dictator* (New York and London: Harper and Brothers, 1934).

21. J. S. Huxley, "TVA, an Achievement of Democratic Planning," *Architectural Review* 93, no. 558 (June 1943): 138–66.

22. J. S. Huxley, *UNESCO: Its Purpose and Its Philosophy* (Washington, DC: Public Affairs Press, 1948).

23. J. S. Huxley, *Religion Without Revelation* (London: Ernest Benn, 1927).

24. Huxley and Huxley, *Evolution and Ethics 1893–1943*, pp. 136–37.

25. J. S. Huxley, *UNESCO*, p. 11.

26. J. S. Huxley, *New Bottles for New Wine* (London: Chatto and Windus, 1957).

27. J. S. Huxley, *Individual in the Animal Kingdom*, pp. 114–15.

28. J. S. Huxley, *Essays of a Humanist* (London: Chatto and Windus, 1964), p. 35.

29. Edward O. Wilson, *On Human Nature* (Cambridge, MA: Harvard University Press, 1978); *Biophilia* (Cambridge, MA: Harvard University Press, 1984); *The Diversity of Life* (Cambridge, MA: Harvard University Press, 1992); *The Creation: A Meeting of Science and Religion* (New York: Norton, 2006).

30. E. O. Wilson, *Naturalist* (Washington, DC: Island Books/Shearwater Books, 1994).

31. C. D. Broad, "Critical Notice of Julian Huxley's *Evolutionary Ethics*" (Reprinted from *Mind* 53 [1944]), in Broad's *Critical Essays in Moral Philosophy*, ed. D. R. Cheney (London: Allen and Unwin, 1971), pp. 185–86.

32. Ibid., p. 186.

33. Ibid., pp. 186–87.

34. G. G. Simpson, *Concession to the Improbable: An Unconventional Autobiography* (New Haven, CT: Yale University Press, 1978); L. Laporte, *George Gaylord Simpson* (New York: Columbia University Press, 2000).

35. J. S. Huxley, *Evolution: The Modern Synthesis*, p. 569.

36. Letter to J. S. Huxley, August 20, 1950, Simpson Papers, American Philosophical Society, Philadelphia.

37. G. G. Simpson, *The Meaning of Evolution* (New Haven, CT: Yale University Press, 1949), p. 311.

38. Ibid., p. 315.

39. Ibid., p. 321.

40. Ibid., p. 324.

41. G. G. Simpson, *This View of Life* (New York: Harcourt, Brace, and World, 1964), p. 143.

42. Ibid., pp. 145–46.

43. J. Murphy, *Evolution, Morality, and the Meaning of Life* (Totowa, NJ: Rowman and Littlefield, 1982), p. 112, n21.

44. M. Ruse and E. O. Wilson, "The Evolution of Morality," *New Scientist* 1478 (1985): 108–28.

45. T. H. Huxley, *Hume* (London: Macmillan, 1879), pp. 239–40.

46. L. Huxley, *The Life and Letters of Thomas Henry Huxley* (London: Macmillan, 1900), 2: 306.

Seven

EVOLUTION AND
THE NOVEL

cholars both of the history of science and of literary theory have long been interested in the connections between evolutionary theory, particularly the evolutionary theory of Charles Darwin as expounded in his *Origin of Species*,[1] and the writings of novelists and other literary figures including poets. There is for instance much interest in the *Origin* itself, and the extent to which Darwin drew on literary conventions and styles of his day. Well known, for instance, is the way in which Darwin frames his arguments in much more of a literary than a formal logical fashion, and the very extensive use that he makes of metaphor. Much effort has been expounded, perhaps not always successfully, to show how Darwin's argumentation reflects the styles of the great Victorian novelists who were writing at his time. And in other respects also there has been interest in the rhetoric of Darwin's writings, particularly for instance as it mirrors or parallels the writings of the natural theologians of his day. I have indeed myself drawn attention to the close similarity between the famous final paragraph of the *Origin of Species* and an equally rhetorical passage penned some twenty years earlier—a passage that we know Darwin studied with some care and interest—by the scientist and natural theologian David Brewster. (See essay 1.)

IN MEMORIAM

Just as scholars have looked at the influence of novels and other writings on the style of Charles Darwin and other evolutionists, so also there has been interest in the way in which literary figures have used evolutionary ideas. Perhaps the most famous use in poetry is one that involves the pre-Darwinian theory of the Scottish evolutionist Robert Chambers, as expressed in his notorious *Vestiges of the Natural History of Creation*.[2] The poet Alfred Tennyson incorporated many of Chambers's ideas into his famous and much-loved poem *In Memoriam*. This poem was begun in the 1830s but not completed until about 1850. It is a testament to the memory of a young friend of Tennyson, Arthur Hallam, whose life was cut short. Tennyson writes at first in the poem about his desolation at Hallam's death and existence's apparent meaninglessness, something that he found reflected in the uniformitarian geology of Charles Lyell. Lyell had argued that nature is going nowhere, just simply bound by unbroken, stern laws, and that there is no end in prospect, nor any progress in view. Life comes and life goes without meaning as expressed in the following famous passage:

> Are God and Nature then at strife,
>   That Nature lends such evil dreams?
>   So careful of the type she seems,
> So careless of the single life;
>
> .   .   .   .
>
> So careful of the type? but no.
>   From scarped cliff and quarried stone
>   She cries, 'A thousand types are gone:
> I care for nothing, all shall go.'[3]

Given Nature "red in tooth and claw"—this is the source of this famous phrase—nothing seems to make any sense. Not only are individuals pointless mortals, but so also are groups. We are born, we live, and then we die—usually painfully. Nothing makes sense or has meaning. There are just endless Lyellian cycles. Then toward the end of the 1840s Tennyson read Chambers, or at least he read a very detailed review of Chambers's *Vestiges*. Chambers argued for an organic evolution that was unambiguously progressionist, that is to say moving up from simple forms up to humans, and then perhaps beyond. Inspired by this, Tennyson picked up pen and finished his poem. He argued in the final lines that perhaps there is meaning after all, despite a Lyellian uniformitarianism: that life is progressing upward, and that perhaps will go on beyond the human form that we have at present. Could it not be that Hallam represented some anticipation of the more-developed life to come, cut short as it were in its prime? There is therefore hope for us all and a meaning for the life of Hallam:

> A soul shall draw from out the vast
> And strike his being into bounds,
> And moved thro' life of lower phase,
>   Result in man, be born and think,
>   And act and love, a closer link
> Betwixt us and the crowning race
>
> .   .   .   .
>
> Whereof the man, that with me trod
>   This planet, was a noble type
>   Appearing ere the times were ripe,
> That friend of mine who lives in God.[4]

DARWIN OR SPENCER?

What about writing after the *Origin*, in particular what about the possible influence of evolutionary ideas on the novel? No one would deny that there was a major influence, although (expectedly) many would argue that the influence was as (or more) often Spencerian than Darwinian. Certainly when one looks at American novelists, around the end of the nineteenth century, one gets themes of inevitability and progress that fit more with the Synthetic Philosophy than with the *Origin of Species*. Look for example at *The Octopus*, a novel penned by Frank Norris, with the theme being the struggle in California between the wheat growers and the all-encroaching railroad company. No one seems to have much control over their individual lives, neither the hero (Presley) or the baron of the railroad (Shelgrim). The latter is ruthlessly forcing the tracks across the state, irrespective of the happiness of others (represented below by the farmer, Derrick). But he is no less a creature of fate and social progress:

"And," continued the President of the P. and S. W. with grave intensity, looking at Presley keenly, "I suppose you believe I am a grand old rascal."

"I believe," answered Presley, "I am persuaded———" He hesitated, searching for his words.

"Believe this, young man," exclaimed Shelgrim, laying a thick powerful forefinger on the table to emphasize his words, "try to believe this—to begin with—*that railroads build themselves*. Where there is a demand sooner or later there will be a supply. Mr. Derrick, does he grow his wheat? The Wheat grows itself. What does he count for? Does he supply the force? What do I count for? Do I build the Railroad? You are dealing with forces, young man, when you speak of Wheat and the Railroads, not with men. There is the Wheat, the supply. It must be carried to feed the People. There is the demand. The Wheat is one force, the Railroad, another, and there is the law that governs them—supply and demand. Men have only little to do in the whole business. Complications may arise, conditions that bear hard on the individual—crush him maybe—*but the wheat will be carried to feed the people* as inevitably as it will grow. If you want to fasten the blame of the affair at Los Muertos on any one person, you will make a mistake. Blame conditions, not men."

"But—but," faltered Presley, "you are the head, you control the road."

"You are a very young man. Control the road! Can I stop it? I can go into bankruptcy if you like. But otherwise if I run my road, as a business proposition, I can do nothing. I can not control it. It is a force born out of certain conditions, and I—no man—can stop it or control it. Can your Mr. Derrick stop the Wheat growing? He can burn his crop, or he can give it away, or sell it for a cent a bushel—just as I could go into bankruptcy—but otherwise his Wheat must grow. Can any one stop the Wheat? Well, then no more can I stop the Road."[5]

This Spencerian philosophy permeates the novel, and just in case you are not quite bright enough to articulate it for yourself, at the end of the novel the author does it for you. Lives have been broken—people are crushed to death by poverty, crime, insanity, prostitution. But the railroad is built and the wheat grows. All for the ultimate benefit of humankind:

Falseness dies; injustice and oppression in the end of everything fade and vanish away. Greed, cruelty, selfishness, and inhumanity are short-lived; the individual suffers but the race goes on. Annixter [a character in the novel] dies, but in a distant corner of the world a thousand lives are saved. The larger view always and through all shams, all wicked-nesses, discovers the Truth that will, in the end, prevail, and all things, surely, inevitably, resistlessly work together for good.[6]

No matter what any individual may do, nothing can stop the march of progress.

In fact, interestingly, the railway chief—Shelgram—is presented as a man with some personal redeeming qualities. He is prepared, for example, to extend the hand of help to a drunken worker. No such soft-ness clouds the vision of the author who, beyond all others, epitomizes this kind of writing: Jack London. His most popular novel, *The Call of the Wild*—checking Amazon.com I found that there are at least thirteen edi-tions now in print, not to mention numerous student cribs and teachers' guides—is Social Darwinian to the core, perhaps expressing a philosophy truly found in neither Spencer nor Darwin but emblematic of the busi-ness philosophy of successful American robber barons of the day. Here, for instance, are the emotions of the hero, the dog Buck, on chasing a rabbit—a rabbit!

> All that stirring of old instincts which at stated periods drives men out from the sounding cities to forest and plain to kill things by chemically propelled leaden pellets, the blood lust, the joy to kill—all this was Buck's, only it was infinitely more intimate. He was ranging at the head of the pack, running the wild thing down, the living meat, to kill with his own teeth and wash his muzzle to the eyes in warm blood.

> There is an ecstasy that marks the summit of life, and beyond which life cannot rise. And such is the paradox of living, this ecstasy comes when one is most alive, and it comes as a complete forgetfulness that one is alive. This ecstasy, this forgetfulness of living, comes to the artist, caught up and out of himself in a sheet of flame; it comes to the soldier, war-mad on a stricken field and refusing quarter; and it came to Buck, leading the pack, sounding the old wolf-cry, straining after the food that was alive and that fled swiftly before him through the moonlight. He was sounding the deeps of his nature, and of the parts of his nature that were deeper than he, going back into the womb of Time. He was mas-tered by the sheer surging of life, the tidal wave of being, the perfect joy of each separate muscle, joint, and sinew in that it was everything that

was not death, that it was aglow and rampant, expressing itself in movement, flying exultantly under the stars and over the face of dead matter that did not move.[7]

When it comes to Buck's fellow dog, things really get into top gear:

> There was no hope for him. Buck was inexorable. Mercy was a thing reserved for gentler climes. He manoeuvred for the final rush. The circle had tightened till he could feel the breaths of the huskies on his flanks. He could see them, beyond Spitz and to either side, half crouching for the spring, their eyes fixed upon him. A pause seemed to fall. Every animal was motionless as though turned to stone. Only Spitz quivered and bristled as he staggered back and forth, snarling with horrible menace, as though to frighten off impending death. Then Buck sprang in and out; but while he was in, shoulder had at last squarely met shoulder. The dark circle became a dot on the moon-flooded snow as Spitz disappeared from view. Buck stood and looked on, the successful champion, the dominant primordial beast who had made his kill and found it good.[8]

It is the inevitability that is so fascinating. Men and dogs are swept along by the laws of nature and there is nothing anyone can do about things. Perhaps we are but incompletely evolved and freedom will come eventually. But do not hold your breath:

> Among the forces which sweep and play through the universe, untutored man is but a wisp in the wind. Our civilisation is still in a middle stage, scarcely beast, in that it is no longer wholly guided by instinct; scarcely human, in that it is not yet wholly guided by reason. On the tiger no responsibility rests. We see him aligned by nature with the forces of life—he is born into their keeping and without thought he is protected. We see man far removed from the lairs of the jungles, his innate instincts dulled by too near an approach to free-will, his free-will not sufficiently developed to replace his instincts and afford him perfect guidance.[9]

Thus Theodore Dreiser in *Sister Carrie* (1900), a novel that tells the story of a kept woman. "In Carrie—as in how many of our worldlings do they not?—instinct and reason, desire and understanding, were at war for the mastery. She followed whither her craving led. She was as yet more drawn than she drew."[10]

Of course there were those who opposed this whole line of thinking. The Irish playwright George Bernard Shaw was one. In the introduction to his play *Back to Methuselah*, Shaw wrote of Darwin and his friends:

> I really do not wish to be abusive; but when I think of these poor little dullards, with their precarious hold of just that corner of evolution that a blackbeetle can understand—with their retinue of twopenny-half-penny Torquemadas wallowing in the infamies of the vivisector's laboratory, and solemnly offering us as epoch-making discoveries their demonstrations that dogs get weaker and die if you give them no food; that intense pain makes mice sweat; and that if you cut off a dog's leg the three-legged dog will have a four-legged puppy, I ask myself what spell has fallen on intelligent and humane men that they allow themselves to be imposed on by this rabble of dolts, blackguards, impostors, quacks, liars, and, worst of all, credulous conscientious fools.[11]

One is glad that he did not wish to be abusive. What Shaw did do was follow up his hatred of the evolutionary biology of his day by coming up with his own, somewhat odd alternative—in the same play (*Back to Methuselah*)—where we are born from eggs, have sex until we are sore, and at four years old lose interest in everything except the life of the mind. His theory was a mishmash of Bergsonian creative evolution and Shavian wit, for one is never quite sure when one is having one's leg pulled.

What I do think becomes more and more obvious, as one reads the words of the novelists (and poets and playwrights) is that to a certain extent the evolution is froth on the top. I do not mean that it plays no part and is irrelevant. *In Memoriam* would not be the same without Vestiges, and *Call of the Wild* would not be the same without some evolutionary theory, whatever it might have been. But although certainly fiction using evolutionary themes often bears a strong connection to what evolutionists were claiming happens in the real world of evolution, it is clear that a lot of the stuff about progress and determinism and fate that finds its way into the fiction bears but a tenuous connection to anything that truly happens in the real world of evolution In reading *Octopus*, for example, one is often reminded of the very similar philosophy expressed by Leo Tolstoy in *War and Peace*. In both novels, the individual counts for virtually nothing. It is all a matter of the inexorable laws of nature. Yet, in Tolstoy it would be hard to imagine anyone further from a philosophy based on evolution.

I suppose in a way one ought to have expected this, but it is worth stating explicitly nevertheless. Evolution may not deserve praise for people's philosophical conclusions, but then neither should it deserve condemnation. Keeping this in mind, let us turn to a more recent piece of fiction, a well-reviewed novel by an award-winning British writer. I am not making the point about the loose connection between evolution and philosophy to say that using evolution in one's writing is a bad thing, or to say that it cannot be very illuminating about human nature. I am saying that I think one should be wary of concluding that the morals drawn in a work of fiction using evolution necessarily stem straight from the evolution. There are probably other factors at play when conclusions are being articulated.

ENDURING LOVE

Enduring Love, by Ian McEwan, published in 1997, draws heavily on contemporary evolutionary ideas, particularly on the so-called sociobiological ideas of writers like the Harvard entomologist and student of social behavior, Edward O. Wilson. The story McEwan tells is as follows. A science writer Joe Rose is having a picnic somewhere in the English countryside with his live-in mistress Clarissa. They see a hot-air balloon in trouble and Joe rushes over to try to help. A number of other people are doing likewise. These other would-be helpers are a young man, Jed Parry; a doctor in his early forties, John Logan; and two others, Joseph Lacey and Toby Greene. Unfortunately the balloon starts to rise up with a small child in the bottom of the basket. All let go except for Logan, who is carried high into the air at the end of a rope. Finally, he falls several hundred feet and is killed.

The story takes off from here because Jed takes an immediate and obsessively erotic interest in Joe. It turns out that Jed is suffering from a known psychological syndrome—de Clérambault's syndrome—that manifests itself by the sufferer falling for another person in an immediate and obsessive way, convinced that this love-object person returns the affection. Although this conviction is a complete delusion, nothing will deter the sufferer from his or her belief. (At the end of the book, McEwan gives a discussion of the syndrome. The syndrome really exists and McEwan's references are nonfictional.)

The main plot of the novel involves Jed's obsession for Joe and the consequences therefrom. At first, Jed phones Joe. Then, Jed starts sending letters and hanging around outside Joe's house. Initially, Joe just brushes this off and does not tell Clarissa. Then, when Joe does let on, he has already started to build suspicion in her mind. Does Jed's passion really exist, or is Joe making this all up? Then, when Clarissa is finally convinced that the passion is genuine, she wonders whether or not it was caused in some sense by Joe, or whether it just appeared from nowhere? Jed gets more and more obsessed, eventually (being unable to get a response from his beloved) hiring a gunman who tries to shoot Joe in the restaurant. The police are unsympathetic to the idea that this might have been brought about by Jed. Hence, Joe (who fears that Jed will try again) goes out and (through an old acquaintance) buys a gun for himself. Jed meanwhile has captured Clarissa in their apartment. Joe arrives back. Jed threatens to commit suicide before them, and Joe shoots him in the arm, thus preventing the fatality from happening. Clarissa now is so upset by the whole occurrence that she leaves Joe. Hitherto, they had had a deep and satisfying relationship, but she feels that she can no longer have the faith in Joe that she once had. However, at the end of the book we are given a hint that perhaps their relationship will rekindle: in the account of the disease on which the book is based it is suggested that Joe and Clarissa come back together again.

Underneath the main story is a subtheme about John Logan, the doctor. Joe is upset because John's death leaves a widow and two small children. He goes to visit the wife. She, however, is desperately upset because there is evidence that Logan (when he went chasing after the balloon) was not where he was supposed to be. She feared that he was about to have a picnic with some unknown woman. (A head scarf was left in Logan's car when he went running over to the balloon.) Eventually, at the end of the story, thanks to the other people who were there at the site of the tragedy (Lacey and Greene), we learn that in fact Logan was entirely innocent. He had been giving a lift to two hitchhikers: a fifty-year-old professor of mathematics and his mistress. These two had good reason to keep their presence at the site anonymous. Logan therefore had not been involved in an extramarital tryst when he died. He was helping others as he was to help the child in the balloon, even unto his own death. It is this act of ultimate sacrifice that suggests at the end that perhaps Joe and Clarissa can come back together: their love rekindled by Logan's selfless action, something that conquers the malignant, evil obsession that so consumes Jed.

EVOLUTIONARY THEMES

Prima facie none of this is particularly evolutionary, but there are many references throughout the book to science (remember, Joe is a science writer). Moreover, at the end of the book, McEwan openly refers to his debt to the writings of evolutionists including those of Edward O. Wilson and others. At the beginning of the book, the topic of evolutionary biology is introduced most directly in the context of discussion of the ideas discussed and generated by Joe in the pursuit of his profession. This in itself is used to point to the fact that evolution will be important: there are many comments about how (today) scientists are thinking deeply not only about evolution but about the evolution of social behavior. There is related talk of how human beings are themselves are products of evolution, and how they work together for evolutionary ends.

Novelists don't just talk about ideas—they try to show them in action. McEwan is no exception. Even before the talk about cooperation gets going, the action has set the scene. Joe and the others are holding onto the balloon and then let go, with the exception of Logan. Why is it, Joe wonders, that they behaved as they did? They were willing to help, yes, but they were not so willing to give up their lives. Hence, with the exception of Logan, out of what one might describe as "selfishness" they let go:

> I didn't know, nor have I ever discovered, who let go first. I'm not prepared to accept that it was me. But everyone claims not to have been first. What is certain is that if we had not broken ranks, our collective weight would have brought the balloon down to earth a quarter of the way down the slope a few seconds later as the gust subsided. But as I've said, there was no team, there was no plan, no agreement to be broken.

No failure. So can we accept that it was right, every man for himself? Were we all happy afterwards that this was a reasonable course? We never had that comfort, for there was a deeper covenant, ancient and automatic, written in our nature. Co-operation—the basis of our earliest hunting successes, the force behind our evolving capacity for language, the glue of our social cohesion. Our misery in the aftermath was proof that we knew we had failed ourselves. But letting go was in our nature too. Selfishness is also written on our hearts. This is our mammalian conflict—what to give to the others, and what to keep for yourself. Treading that line, keeping the others in check and being kept in check by them, is what we call morality. Hanging a few feet above the Chilterns escarpment, our crew enacted morality's ancient, irresolvable dilemma: us, or me.[12]

But, apparently, there is a paradox here because Logan did not let go. Could this be a denial of sociobiology (that branch of evolution dealing with social behavior)? These days, group selection is out of favor. Generally, the modern evolutionist thinks that we all act for selfish ends. We are, in Richard Dawkins's memorable phrase, driven by "selfish genes." Why then did Logan show such apparent unselfishness? The sociobiological dilemma is solved when it appears that Logan had a mistress. Surely, therefore, Logan was showing off in front of this mistress and so in some way was trying to promote himself? His act was directed toward his own self-advancement. The fact that it was to prove fatal was just a contingent unfortunate consequence. Logan was rather like a peacock displaying his tail, with the intention of attracting the peahen. This is true even though sometimes the long tail proves fatal when pursued by a predator.

All of this seems to be just a teaser for what is to come later, although as we shall see this point is going to be woven skillfully into the climax of the novel. The main use of sociobiology by McEwan seems to be over Joe's actions: whether or not the way in which he is behaving is in some sense biological or not. At one level, it is suggested that our understanding of the whole of nature is a sham brought about by our genes acting to make us efficient reproducers. In a way, even the obsession of Jed is explained as being something that is normal or natural from a biological point of view: abhorrent of course, but no less and no more an illusion than the illusion of others. In particular, it is argued, very much in line with the arguments of Edward O. Wilson (given in *On Human*

Nature, one of the books referenced by McEwan), that Jed's religious obsession is on a par with anybody's religious beliefs:

> While I was waiting for the kettle I looked at a radio talk I was going to record that afternoon. I remember it well because I used the material later for the first chapter of a book. Might there be a genetic basis to religious belief, or was it merely refreshing to think so? If faith conferred selective advantage, there were any number of possible means, and nothing could be proven. Suppose religion gave status, especially to its priest caste—plenty of social advantage in that. What if it bestowed strength in adversity, the poser of consolation, the chance of surviving the disaster that might crush a godless man? Perhaps it gave believers passionate conviction, the brute strength of single-mindedness.
>
> Possibly it worked on groups as well as on individuals, bringing cohesion and identity, and a sense that you and your fellows were right, even—or especially—when you were wrong. With God on our side. Uplifted by a crazed unity, armed with horrible certainty, you descend on the neighboring tribe, beat and rape it senseless and come away burning with righteousness and drunk with the very victory your gods had promised. Repeat fifty thousand times over the millennia, and the complex set of genes controlling for groundless conviction could get a strong distribution. I floated in and out of these preoccupations. The kettle boiled and I made the tea.[13]

The suggestion is that life is a kind of illusion. We are all caught in evolution's fancy, save we can break out in some way through science or something of that nature. Certainly, this is an idea repeated later in the book:

> I felt a familiar disappointment. No one could agree on anything. We lived in a mist of half-shared, unreliable perception, and our sense data came warped by a prism of desire and belief, which tilted our memories too. We saw and remembered in our own favor and we persuaded ourselves along the way. Pitiless objectivity, especially about ourselves, was always a doomed social strategy. We're descended from the indignant, passionate tellers of half truths who in order to convince others, simultaneously convinced themselves. Over generations success had winnowed us out, and with success came our defect, carved deep in the genes like ruts in a cart track—when it didn't suit us we couldn't agree on what was in front of us. Believing is seeing. That's why there are

divorces, border disputes and wars, and why this statue of the Virgin Mary weeps blood and that one of Ganesh drinks milk. And that was why metaphysics and science were such courageous enterprises, such startling inventions, bigger than the wheel, bigger than agriculture, human artifacts set right against the grain of human nature. Disinterested truth. But it couldn't save us from ourselves, the ruts were too deep. There could be no private redemption in objectivity.[14]

So, at some level, there seems to be a basic theme that everything is not so much unreal but a kind of dream-world brought on by natural selection. This happens because natural selection is interested in only reproduction rather than in us getting to the objective truth. However, McEwan is more skilled than this. He has a more sophisticated underlying theme. Again he builds on the ideas of the sociobiologists: although it is true that we humans are caught in illusion, perhaps in some sense we humans (uniquely) are able to break loose from this false consciousness. We can transcend our purely biological nature: through such things as science and the like, we can start to discover how life truly and really is, rather than how it just appears to us.

At some later point in the novel, just after Joe has bought the revolver he is going to use eventually to shoot Jed, he is returning to London. Overcome with fear, he feels an immediate need to defecate. He goes into the woods and scrapes out a hole:

> I left him waiting in the front seat while I took some paper and went back into the trees, and used my heel to scrape a shallow trench. While I crouched there with my pants around my ankles, I tried to soothe myself by parting the crackly old leaves and scooping up a handful of soil. Some people find their long perspectives in the stars and galaxies; I prefer the earthbound scale of the biological. I brought my palm close to my face and peered. In the rich black crumbly mulch I saw two black ants, a springtail, and a dark red worm-like creature with a score of pale brown legs. These were the rumbling giants of this lower world, for not far below the threshold of visibility was the seething world of the roundworms—the scavengers and the predators who fed on them, and even these were giants relative to the inhabitants of the microscopic realm, the parasitic fungi and the bacteria—perhaps ten million of them in this handful of soil. The blind compulsion of these organisms to consume and excrete made possible the richness of the soil, and therefore

the plants, the trees, and the creatures that lived among them, whose number had once included ourselves. What I thought might calm me was the reminder that, for all our concerns, we were still part of this natural dependency—for the animals that we ate grazed the plants which, like our vegetables and fruits, were nourished by the soil formed by these organisms. But even as I squatted to enrich the forest floor, I could not believe in the primary significance of these grand cycles. Just beyond the oxygen-exhaling trees stood my poison-exuding vehicle, inside which was my gun, and thirty-miles down teeming roads was the enormous city on whose northern side was my apartment where a madman was waiting, a de Clérambault, my de Clérambault, and my threatened loved one. What, in this description, was necessary to the carbon cycle, or the fixing of nitrogen? We were no longer in the great chain. It was our own complexity that had expelled us from the Garden. We were in a mess of our own unmaking. I stood and buckled my belt and then, with the diligence of a household cat, kicked the soil back into my trench.[15]

Now why is this relevant? What McEwan suggests is that Joe, through his knowledge and love of science, has managed in some sense to transcend his purely biological nature. But the paradox seems to be that, inasmuch and because he has transcended his biological nature, Joe only makes things worse: for himself, for Jed, and particularly for the relationship between himself and Clarissa. This comes because Joe has been able to see through Jed, to see the way that his mind is working and that he has got an illness, and because Joe wants to control the situation. Unfortunately, if anything, things become worse. The police do not believe him. Clarissa is mad and hurt by him. (At one point, Joe goes through Clarissa's mail trying to find evidence that she is turning against him because of other relationships. She finds out with predictable emotional consequences.) And, expectedly, Joe's going out and finding a gun to shoot Jed, rather than leaving it to others, upsets everyone. In the "Dear Joe" letter that Clarissa writes, saying that she is leaving him, she makes it very clear that it is precisely because Joe would not stay at the half-world of illusion that things have soured between them. He would not leave things at the familiar level, but insisted on digging down to find the truth, whatever that might be. As a result, their relationship must come to an end:

You went your own way, you denied him everything, and that allowed his fantasies, and ultimately his hatred, to flourish. You asked me last night if I realized that you had saved my life. In the immediate sense, of course, that's true. I'll always be grateful. You were brave and resourceful. In fact you were brilliant. But I don't accept that it was always inevitable that Parry was going to hire killers or that I should end up being threatened with a knife. My guess was that he was always more likely to do himself harm. How wrong and how right I was! You saved my life, but perhaps you put my life in jeopardy—by drawing Parry in, by overreacting all along the way, by guessing his every next move as if you were pushing him towards it.

A stranger invaded our lives, and the first thing that happened was that you became a stranger to me. You worked out he had de Clérambault's syndrome (if that really is a disease) and you guessed he might become violent. You were right, you acted decisively and you're right to take pride in that. But what about the rest?—why it happened, how it changed you, how it might have been otherwise, what it did to us— that's what we've got now, and that's what we have to think about.[16]

Apparently, in the end, McEwan is offering a very depressing novel: if we try to break out of our biological nature, then precisely because of our biology, this is only to lead to great unhappiness. As Wilson suggests, there are very good biological reasons for our illusory world. Better by far for us to remain down at the animal level. But McEwan is too skilled a novelist to leave things at this level. Finally, he comes back tying in both his main story and his subtheme. He picks up again on the beginning passage about the men hanging onto the balloon and about only Logan staying true to the course. Now we know that it was untrue that Logan's apparent altruism was bogus and that he was selfishly trying to impress his mistress. There was no mistress. Or rather, the scarf belonged to the mistress of someone else. Logan's behavior was therefore purely disinterested. He went over to help the boy in the balloon in distress purely for noble reasons.

In other words, Logan rose above his own biological selfishness. In the words of the professor who was being given a lift by Logan: "He was a terribly brave man. . . . It's the kind of courage the rest of us can only dream about."[17] So, in the end, what we see is that Logan has managed to transcend his biological nature, just as Joe transcended his nature. Yet, whereas Joe's move seems to have plunged us into despair and destruction, Logan's selflessness—his break from his biological nature—pushes us up to a higher level. And, as I have said, it is suggested in the appendix discussing the clinical case of de Clérambault's syndrome that this is enough to overcome the harm that Joe has done, and that in the future Joe and Clarissa will come back together and by adopting a child have a family. (Part of the theme of Joe and Clarissa's love rests on the fact that Clarissa is unable to have children. They are trying to create a love without children.)

DETERMINISM

It goes without saying that McEwan's view of life resonates far more with the beginning of the twenty-first century than either Jack London's "nature red in tooth and claw" view of things (the "law of club or fang" is London's phrase), or the fatalistic determinism of Frank Norris and Theodore Dreiser—or, I trust, than the fantasies of George Bernard Shaw. In part this is because the science has moved on. Nature red in tooth and claw was never truly Darwinian (as we saw, the phrase was Tennyson's) and now today, thanks to studies of social behavior, we do realize how fully cooperation exists in nature and how vital it is and (despite disagreements about levels of selection) how this is a product of natural selection. In part, it is because McEwan's philosophy is more in tune with modern times. After the horrors of the twentieth century, simplistic assumptions about progress are no longer popular, and neither is the claim that none of us have ultimate responsibility for how nature and society takes its course. That was one of the arguments of the defendants at the Nuremberg Trials and too often heard from the mouths of others who claim that what happens in this life is not their responsibility.

However, referring back to the point made earlier, do note that McEwan's solution does not stem from the biology. His way of tackling the issue of the doctor's extreme altruism is to take the situation out of the biological. We have already been softened up for this move and now it is made in full. Dr. Logan is a truly good man—shows real love— because he has escaped the chains of the genes. At the final crucial moment, he is no longer *Homo biologicus*. He is the person making sense of the novel's title, the person of whom Saint Paul was speaking. (Love "endures all things," 1 Corinthians 13:7.) And my point is that although this may be the right way, it is not a way dictated by the nature of modern evolutionary biology. There are many who like this way. Richard Dawkins would seem to be one. At the end of *The Selfish Gene* he argues that we can and must escape from the tyranny of the genes. Marxist biologists like Richard Lewontin and Stephen Jay Gould—forever declaiming against the sins of biological determinism—are others. And there are many philosophers and religious folk who feel the same way.

But on the other side, there are those (I am one) who think that biology might be able to explain even cases of extreme altruism like Logan's. We do not think it necessarily beyond the power of reciprocal altruism to produce even the most extreme cases of help to others. At least, we think it is worth trying to make such a case without dismissing it at once as impossible. More than this, there is a venerable philosophical tradition—David Hume being its most distinguished exponent—that argues that freedom is compatible with laws, including biological laws. That argues in fact that freedom demands laws because without laws you just have randomness and chaos—you certainly do not have moral responsibility. This being so, it would certainly be open to a skilled novelist to argue that, far from Logan's goodness stemming from the fact that somehow he transcended his human nature, he was good precisely because he was obeying his human nature. Outside human nature lies monsterdom, not sainthood. I am not saying that a novelist should argue this. I am saying a novelist could argue this. And this being so, I am concluding what I argued before, namely, that one should be wary of taking the endings of novels at face value. Good tellers of tales cleverly weave the threads of evolution into their fabrics, but you should not think that the finished product owes everything to these threads—the strands are but part of the whole cloth.

POSTSCRIPT

I have been discussing fiction in this essay. In writing it, I have become more and more aware that the chief point I am trying to make—that the philosophies of the writers are at least as important as the ideas of science on which they rely—apply no less to nonfiction. Today, a very popular cry in certain evangelical Christian circles is that Darwinism is evil because it led to Hitler and the ideas in *Mein Kampf*. (See for example Weikart's *From Darwin to Hitler*.)[18]My response is that even if there are indeed evolutionary ideas in the National Socialist Philosophy—and that in itself needs arguing—we should be very wary of concluding that the Nazi ideology was therefore in some genuine sense Darwinian or some such thing. At best it may be Darwinian in the sense that *Call of the Wild* is Darwinian, and that, in my opinion, is not very much.

NOTES

1. C. Darwin, *On the Origin of Species by Means of Natural Selection, or the Preservation of Favoured Races in the Struggle for Life* (London: John Murray, 1859).

2. R. Chambers, *Vestiges of the Natural History of Creation* (London: Churchill, 1844).

3. Alfred, Lord Tennyson, *In Memoriam*, canto 55, 56, in *In Memoriam: An Authoritative Text, Backgrounds and Sources Criticism*, ed. R. H. Ross (New York: Norton, 1974), pp. 3–90.

4. Ibid., epilogue.

5. F. Norris, *Octopus: A Story of California* (New York: Doubleday, Page, 1901), p. 576.

6. Ibid., pp. 651–52.

7. J. London, *The Call of the Wild* (New York: Macmillan, 1903), chapter 3: "The Dominant Primordial Beast."

8. Ibid.

9. T. Dreiser, *Sister Carrie* (New York: Doubleday, Page, 1900), chapter 8: "Intimations by Winter—An Ambassador Summoned."

10. Ibid.

11. G. B. Shaw, *Back to Methuselah: A Metabiological Pentateuch* (Harmondsworth, UK: Penguin, 1921).

12. I. McEwan, *Enduring Love* (London: Cape, 1997), pp. 14–15.

13. Ibid., p. 159.

14. Ibid., pp. 180–81.

15. Ibid., pp. 206–207.

16. Ibid., p. 218.

17. Ibid., p. 230.

18. R. Weikart, *From Darwin to Hitler: Evolutionary Ethics, Eugenics, and Racism in Germany* (New York: Palgrave Macmillan, 2004).

PART IV:
THE LATER YEARS

*A*mong the big mistakes made by the Creationists—and there is strong competition to get into this set—is the belief that if a scientific theory does not yield all of the answers, at once, it must be seriously deficient. This could not be further from the truth. The mark of really great science—think of Newtonian mechanics—is that it gives future workers masses of really interesting problems to tackle and to solve. No one wants to sit around for a hundred years polishing the double helix. That was great science because, before Watson and Crick had even finished their paper, they were speculating on the genetic code and the challenge of cracking it. This is one of the reasons why, in the preface to the last section, I expressed some ambivalence about Wright's shifting balance theory. It was wrong, dead wrong, but it led to fifty years of exciting research. That is no small achievement.

The great thing about Darwin's theory of evolution through natural selection is that there is as much work to be done now as there was back in 1859. More even. And one of the areas today that is most exciting is what used to be called embryology, but now goes under the term of development, or more fully "evolutionary development" or *evo-devo* for short. The first essay of this last section looks at evo-devo—its great triumphs and also the ambivalence expressed by some of its enthusiasts about its connection with conventional (that is, Darwinian) evolutionary theorizing. My own feeling is that, were he alive today, there would be no greater supporter of evo-devo than Charles Robert Darwin. This leads me to suspect that something is afoot here (as Sherlock Holmes would have said), and my claim is that there is indeed something going on here and that (as always) in order to understand things we must put them in broader context and go back into the past. I suggest that biological thinking generally has been torn by two different paradigms or metaphors, and that in a sense they are more basic even than the coming of evolutionary thinking. Once these paradigms or metaphors are uncovered, a lot that seems puzzling is puzzling no longer.

Essay 9, the middle essay of this section, looks at recent work attempting to give an evolutionary explanation of religion. Naturalistic accounts of religion go back to the Greeks. Most famously in recent times was David Hume's *Natural History of Religion*, a work I suggest influenced Charles Darwin when he came to write on religion. Note that although Darwin was versed in philosophy and had some interest in its problems, he always writes as a scientist. He is interested in explaining things, not in justifying them. Since Darwin, the naturalistic approach to religion has continued, although perhaps expectedly in an age when both Freud and the social sciences flourished, the main explanations were more in terms of social and personal factors than purely biological. However, in recent years, particularly with the rise of human sociobiology or evolutionary psychology, there has been renewed interest in the Darwinian approach of offering biological explanations of religious beliefs and movements. This essay covers some of the most prominent attempts in this direction, ending bemused at the variety of different suggestions! It seems to me almost a given that something as significant to human beings as religion has to have some connection with evolution. It is equally given that, in light of the fact that, even if one religion is true or if many religions have grasped elements of the truth, not every religion could possibly be true through and through—one or other of the Catholics or the Mormons has to be wrong somewhere—and that surely cries out for some kind of understanding that includes evolution. So the aim of the essay is not to claim that evolution, Darwinian evolution, is irrelevant. It is to say that we have a way to go yet. As you will infer from my comments made about the first essay in this section, I do not necessary view this conclusion with despair and gloom.

Finally, completing the section and the collection, I turn directly to a topic that has certainly been hinted at in earlier essays. I have long thought that there is something a little fishy about many of the things that are said in the name of evolution—often in the name of Charles Darwin, although my suspicion is even more often under the influence of Herbert Spencer. Creationists claim that evolution is a religion, and I am inclined to agree with them. At once I qualify this by saying that I think Creationism is nothing but a religion and Darwinian thinking is much more than a religion. There is an area of professional science, Darwinian evolutionary studies, that is science pure and simple. But there always has been, long before Darwin, a side to evolutionary thought that is a kind of secular humanism, a Christianity substitute. Sometimes the practitioners have been explicit: Julian Huxley is one. Edward O. Wilson is another. Sometimes the practitioners have been anything but explicit. They are in denial. I argue that Richard Dawkins is such a person. My claim is not that it is necessarily wrong to make your evolution into a religion, although it is not something that I very much want to do, but that we should recognize what is going on. And when we want to claim that something is scientific then it should be scientific, and when we want to claim something that is not scientific then we should not pretend that it is what it is not. To do otherwise is to discredit one of the greatest ideas of all time, Darwinian evolutionary theory.

Eight

EVO-DEVO

A New Evolutionary Paradigm?

The homologies of process within morphogenetic fields provide some of the best evidence for evolution—just as skeletal and organ homologies did earlier. Thus, the evidence for evolution is better than ever. The role of natural selection in evolution, however, is seen to play less an important role. It is merely a filter for unsuccessful morphologies generated by development. Population genetics is destined to change if it is not to become as irrelevant to evolution as Newtonian mechanics is to contemporary physics.

—S. F. Gilbert, J. M. Opitz, and R. A. Raff,
"Resynthesizing Evolutionary and Developmental Biology"

These are exciting days for evolutionary biology. In the past twenty years or so, the molecular approach to biology—evolutionary development, or more familiarly, "evo-devo"—has swept all before it. Now we can trace development from the gene to the finished organism. Along the way, some magnificent discoveries have been made, most significantly that organisms as different as the fruit fly and the human share hugely important genes for development.[1] We humans are put together in the same way as are those little insects that hang around compost heaps and rotting vegetables in garbage cans. This has led some enthusiasts to think that we are on the verge of—certainly in need of—a whole new theory of evolution. A new paradigm, that rejects Darwinian natural selection, or that at least reduces it to an unimportant role in cleaning up after the real work has been done. Now we have or are after a new theory—perhaps one that makes the really creative work appear in the course of development. Nature unfurls according to its molecularly based developmental laws, and that is where the real sources of change should be sought. It is true that the fittest survive, but this is little more than a truism with no real evolutionary import.

I am not sure that this is so. I applaud the new work in evo-devo—I think it is some of the most exciting that has ever been done by evolutionists. But whether this spells the demise of natural selection—the end of Darwinian evolutionary theory as we know it—is altogether another matter. I think, in fact, we are seeing an ongoing debate that dates back to Aristotle, between what biologists have labeled "form" and "function." The debate predates the coming of evolution, and it postdates it also[2]—which has interesting consequences for those of us trying, as philosophers, to understand the nature of science. In a way, I do think we have different paradigms, but unlike the normal understanding of "paradigm," I am not sure that one is replacing—will ever replace—the other. This is something that biologists should realize and philosophers should try to understand. Which is why I am writing this essay.

FORM AND FUNCTION

We have two ways of looking at organisms. Although I admit fully that there are differences from his usage, because there is convergence between what is going on in biology with the usage of that term by Thomas Kuhn in his *Structure of Scientific Revolutions*, it is useful and illuminating to speak of them as two different *paradigms*. For Kuhn, a paradigm is a way of looking at things—now you see it one way, now you see it another. Now you see the Earth stationary and everything going around it; now you see the Sun stationary and everything going around it. This is not just a question of reason but more a question of attitude. From taking one or the other position, a whole discipline can be founded, with its own research problems or puzzles and its own techniques and much more.

It is in this sense that, in this essay, I use the word *paradigm*. In the world of organisms, you either see function as dominant—as all-pervasive—and everything else as secondary, or you do not. If you do not, then various formal structures and so forth are what dominate your thinking. If you are a functionalist, then your job is to find function and explain it. If you are a formalist, then your job is to find structure and explain it. Differing from a Kuhnian paradigm, no one is exactly and exclusively a subscriber to the one position and not to the other—although some people get pretty close to the extremes—and these positions and commitments are ongoing. As noted, the psychological picture—the way of seeing—is what is central to a Kuhnian paradigm, and it is this that is central to the way in which I am using "paradigm" here. Some see function, and form is secondary and virtually a nuisance, and some see form, and function is secondary and virtually a nascence.

Aristotle, a first-class biologist, spotted both senses or ways of viewing organisms. On the one hand we have function, or what today is often called *teleology*. What Aristotle called *final cause*:

> [We have cause] in the sense of end or that for the sake of which a thing is done, e.g., health is the cause of walking about. ('Why is he walking about?' We say: 'To be healthy,' and, having said that, we think we have assigned the cause.) The same is true also of all the intermediate steps which are brought about through the action of something else as means towards the end, e.g., reduction of flesh, purging, drugs, or surgical instruments are means towards health. All these things are for the sake of the end, though they differ from one another in that some are activities, others instruments.[3]

Aristotle chided, "What are the forces by which the hand or the body was fashioned into its shape?" A woodcarver (speaking of a model) might say that it was made as it is by tools like an axe or an auger. But note that simply referring to the tools and their effects is not enough. One must bring in ends. The woodcarver "must state the reasons why he struck his blow in such a way as to effect this, and for the sake of what he did so; namely, that the piece of wood should develop eventually into this or that shape." Likewise against the physiologists, "the true method is to state what the characters are that distinguish the animal—to explain what it is and what are its qualities—and to deal after the same fashion with its several parts; in fact, to proceed in exactly the same way as we should do, were we dealing with the form of a couch."[4]

On the other hand, there is form: "Whatever parts men have in front, these parts quadrupeds have below, on the belly; and whatever parts men have behind, these parts quadrupeds have on their back."[5] In the case of organisms, the similarities or isomorphisms between the parts of very different animals—isomorphisms between parts with very different functions. Traditionally, one refers here to the forelimbs of vertebrates being the favorite example. The arm of man, the wing of bird, the front leg of horse, the flipper of whale, the paw of mole—all used for very different purposes but with similarities in their structures. There seems to be no final cause at work here. Homologies do not exist for the sake of some end. (We today use the word *homology*, meaning similarity due to common descent. Aristotle was not an evolutionist. But, although the word was not first applied until the 1840s, for clarity we can use the term without undue anachronism.)

FUNCTION ACROSS THE EVOLUTIONARY DIVIDE

The important point from our perspective is that these two ways of looking at organisms go across the evolutionary divide. There are pre-evolutionary functionalists. There are postevolutionary functionalists. There are pre-evolutionary formalists. There are postevolutionary formalists. To take the former, consider the famous arguments of the clergyman-naturalist John Ray (1628–1705), especially in his *Wisdom of God, Manifested in the Words of Creation*.[6] His way of thinking is functional through and through:

> Whatever is natural, beheld through [the microscope] appears exquisitely formed, and adorned with all imaginable Elegancy and Beauty. There are such inimitable gildings in the smallest Seeds of Plants, but especially in the parts of Animals, in the Lead or Eye of a small Fry; Such accuracy, Order and Symmetry in the frame of the most minute Creatures, a Louse, for example, or a Mite, as no man were able to conceive without seeming of them.

Everything that we humans do and produce is just crude and amateurish compared to what we find in nature. But this is not a conclusion that is then used to prove evolution. The very opposite. It is used for the argument to design. The living world was likened to a product of design. A machine implies an architect or an engineer, and so likewise inasmuch as the world of life is machine-like, it, too, implies a being, as much above us as the world of life is above our artifacts and creations:

> There is no greater, at least no more palpable and convincing argument of the Existence of a Deity, than the admirable Art and Wisdom that discovers itself in the Make and Constitution, the Order and Disposition, the Ends and uses of all the parts and members of this stately fabric of Heaven and Earth.[7]

A neat package, as the teleological way of thought in biology was tied back into the proof of the divine—the classic argument for design:

> That under one skin there should be such infinite variety of parts, variously mingled, hard with soft, fluid with fixt, solid with hollow, those in rest with those in motion:—all these so packed and thrust so close together, that there is no unnecessary vacuity in the whole body, and yet so far from clashing or interfering with one another, or hindering each others motions, that they do all help and assist mutually on the other, all concur in one general end and design.[8]

Moreover, this is design that is of absolutely the top quality, and so the same must be said of the intelligence behind it. This points to a being worthy of worship, not to some ethereal, local spirit before which the heathen humble themselves.

Then, on the other side of the evolutionary divide, we have Charles Darwin himself. Function is the starting point of his thinking about organisms. Take the little book on orchids, which Darwin published just after the *Origin*. There Darwin was laying out evolutionary biology as he hoped it would be done: "I think this little book will do good to the Origin, as it will show that I have worked hard at details, and it will perhaps, serve [to] illustrate how natural History may be worked under the belief of the modification of species."[9] It is teleological through and through. The very title flags you to this fact: *On the Various Contrivances by which British and Foreign Orchids are Fertilized by Insects, and on the Good Effects of Intercrossing.* It flags you also to the fact that, as with Paley, Darwin was looking at the organic world as if it were an object of design: he was taking organized, end-directed complexity as the absolutely crucial key to unlocking the secrets of the living world and its attributes. Contrivances are human-made objects, which are created with an end in view, as in: "I have invented a remarkable contrivance for shelling hot chestnuts without burning your fingers." This was Darwin's perspective on the living world, just as it had been for Paley.

Then, when we get into the text of the orchids book, the natural theological perspective—the argument to complexity, that is—was used constantly, with great effect. Thus right at the beginning, speaking of how an orchid is fertilized, Darwin described in detail the "complex mechanism" that causes this to happen. There are little sacks of pollen that are brushed by an insect as it pushes its way in, in search of nectar. But not just little sacks. Rather, little sacks (or balls) that are going to go traveling:

> So viscid are these balls that whatever they touch they firmly stick to. Moreover the viscid matter has the peculiar chemical quality of setting, like a cement, hard and dry in a few minutes' time. As the anther cells are open in front, when the insect withdraws its head, . . . one pollinium, or both, will be withdrawn, firmly cemented to the object, projecting up like horns.[10]

Then when the insect visits another plant, the pollen is transferred. But not just by chance:

> How then can the flower be fertilised? This is effected by a beautiful contrivance: though the viscid surface remains immoveably affixed, the apparently insignificant and minute disc of membrane to which the caudicle adheres is endowed with a remarkable power of contraction . . . , which causes the pollinium to sweep through about 90 degrees, always in one direction, viz., towards the apex of the proboscis . . . , in the course, on an average, of thirty seconds.[11]

And so on and so forth. Right through the book, the picture was one of complexity, of adaptation, of function, of purpose:

When we consider the unusual and perfectly adapted length, as well as the remarkable thinness, of the caudicles of the pollinia; when we see that the anther cells naturally open, and that the masses of pollen, from their weight, slowly fall down to the exact level of the stigmatic surface, and are there made to vibrate to and fro by the slightest breath of wind till the stigma is struck; it is impossible to doubt that these points of structure and function, which occur in no other British Orchid, are specially adapted for self fertilization.[12]

And:

In many Vandeæ the caudicles are easily ruptured, and the fertilisation of the flower, as far as this point is concerned, is a simple affair; but in other cases the strength of the caudicles and the length to which they can be stretched before they break is surprising. I was at first perplexed to understand what good purpose the great strength of the caudicles and their capacity of extension could serve.[13]

And of course, this functional perspective on nature was just what natural selection was supposed to address. Selection implies not just change but change toward adaptation, contrivance, function. First the struggle: "A struggle for existence inevitably follows from the high rate at which all organic beings tend to increase."[14] And then on to natural selection. Can it "be thought improbable, seeing that variations useful to man have undoubtedly occurred, that other variations useful in some way to each being in the great and complex battle of life, should sometimes occur in the course of thousands of generations?" Of course this is what happens! "This preservation of favourable variations and the rejection of injurious variations, I call Natural Selection."[15] (See essay 5 for these passages in full.)

FORM ACROSS THE EVOLUTIONARY DIVIDE

Let us turn to the other side, that of formalism. Pre-evolutionary formalists included German thinkers like Johann Wolfgang von Goethe, the great poet; French morphologists like Étienne Geoffroy Saint-Hilaire; as well as British thinkers.[16] The classic case was Richard Owen, who saw all organisms as being based on one fundamental archetype: "What Plato would have called the 'Divine idea' on which the osseus frame of all vertebrate animals . . . has been constructed."[17] Postevolutionary formalists abounded at the time of Darwin. In fact, it was the author of the *Origin* who was the odd man out. For all that he was a great critic of Richard Owen, Thomas Henry Huxley worked exclusively in terms of types. In the textbook he coauthored with his student H. N. Martin, Huxley appealed to the type as he led the students through the living world group by group: mussels, snails, lobsters, frogs, and so forth.[18] The same was true of others in Britain. The brilliant embryologist Frank Balfour wrote that, with respect to the rabbit, "there are grounds for thinking that not inconsiderable variations are likely to be met with in other species,"[19] but this did not stop him from using that very animal as the model when he discussed the mammals, saying indeed that for "the early stages the rabbit necessarily serves as type."[20]

Likewise in Germany, notwithstanding his leadership of the "Darwinismus" movement, Ernst Haeckel owed far more to Goethe and other transcendentalists than to anything in England. The same was true also of his some-time co-worker, the morphologist Carl Gegenbaur. The latter wrote explicitly of "the subordinate importance we must assign to the physiological duties [i.e., functions] of an organ when we are engaged in an investigation in Comparative Anatomy."[21] He was open in his judgment that "physiological value" was secondary and only to be used to avoid mistakes in comparisons. And in America we find Louis Agassiz, student of the *Naturphilosophen* Lorenz Oken and Friedrich Schelling, arguing that "one single idea has presided over the development of the whole class, and that all the deviations lead back to a primary plan."[22] Unlike the teacher, Agassiz's most brilliant student, the invertebrate paleontologist Alpheus Hyatt, became an evolutionist at some point in the 1860s, a Lamarckian of some kind. Yet, so much had he absorbed his master's teachings, that from his papers no one can quite tell when Hyatt made the transition! At some important level, evolution really did not matter. And function was a positive handicap to the real work at hand.[23]

Hyatt was not alone. Even when people thought about change, adaptation and function were downplayed. Although *Naturphilosophie* and its legacy was developmental through and through, it generated an attitude where the type rather than the end was the fundamental organizing concept. The model here, of course, is the individual organism—the very idea of evolution itself is linked to individual growth. Hence, there was a long tradition of drawing an analogy between the order in which organisms appear in the fossil record and the order of the various changes in the embryonic development of the single organism.[24] Agassiz indeed drew a threefold parallelism between the history of life, the history of the individual, and the order of complexity to be found across today's organisms; "One may consider it as henceforth proved that the embryo of the fish during its development, the class of fishes as it at present exists in its numerous families, and the type of fish in its planetary history, exhibit analogous phases through which one may follow the same creative thought like a guiding thread in the study of the connection between organized beings."[25]

With the coming of evolution, Haeckel converted the history into phylogeny—the history of the developing class—and thus he had his so-called biogenetic law, that the history of life can be found in the history of the organism: "Ontogeny recapitulates phylogeny." In fact, morphologists themselves grew increasingly critical of this law in its crude form—there are far too many exceptions—but generally speaking, however people drew the connections between paleontology and embryology, there was little or no interest in seeing the development of the type through time as being selection driven. Other forces, internal forces akin to the forces that lead to individual growth, were thought the key to organic history.

CONSTRAINT

As is well known, in the 1930s the world of evolution took a dramatic turn to Darwinism and function. Thanks to the work of the population geneticists like R. A. Fisher and J. B. S. Haldane in Britain and Sewall Wright in America, and then the experimentalists and empiricists like E. B. Ford in Britain and Theodosius Dobzhansky in America, natural selection was recognized as the dominant mechanism of evolutionary change, and function was seen to be the dominant feature of living beings.[26] Entirely typical of this ultra-Darwinian—ultrafunctionalist—

attitude today is the British biologist and popular science writer Richard Dawkins. He asks what job we expect an evolutionary theory to perform:

> The answer may be different for different people. Some biologists, for instance, get excited about "the species problem," while I have never mustered much enthusiasm for it as a "mystery of mysteries." For some, the main thing that any theory of evolution has to explain is the diversity of life-cladogenesis [branching like a tree]. Others may require of their theory an explanation of the observed changes in the molecular constitution of the genome. I would not presume to try to convert any of these people to my point of view. All I can do is to make my point of view clear, so that the rest of my argument is clear.[27]

Dawkins goes on to agree with John Maynard Smith that "the main task of any theory of evolution is to explain adaptive complexity, i.e. to explain the same set of facts which Paley used as evidence of a Creator." Jokingly he refers to himself as a "neo-Paleyist," concurring with the natural theologian "that adaptive complexity demands a very special kind of explanation: either a Designer as Paley taught, or something such as natural selection that does the job of a designer. Indeed, adaptive complexity is probably the best diagnostic of the presence of life itself."[28]

But now we have this position challenged. The mantra seems to be "constraint." Because of constraints, supposedly natural selection cannot do its job and function is not realized. John Maynard Smith (writing with a group of coauthors) tells us what is at issue: "Organisms are capable of an enormous range of adaptive responses to environmental challenge. One factor influencing the pathway actually taken is the relative ease of achieving the available alternatives. By biasing the likelihood of entering onto one pathway rather than another, a developmental constraint can affect the evolutionary outcome even when it does not strictly preclude an alternative outcome."[29] And, we are told, the study of evolutionary development reveals many constraints. Hence we cannot rely on natural selection, function is to be downgraded, and other more formal factors are to come to the fore.

GENETIC CONSTRAINTS

What about some examples of constraints in action? One much-cited set of examples focuses on so-called genetic constraints. The developmental morphologist Rudolf Raff invites us to look at the issue of genome size. "Having a large genome has consequences outside of the properties of the genome per se. Larger genomes result in larger cells. Because cells containing large genomes replicate their DNA more slowly than cells with a lower DNA content, large genomes might constrain organismal growth rates. Cell size will also determine the cell surface-to-volume ratio, which can affect metabolic rates."[30] Salamanders are one kind of organism that often have large genome sizes. Hence, they seem to be good organisms on which to test hypotheses about constraints, and it seems that the formalists do have a point.

> Roth and co-workers have observed that in both frogs and salamanders, larger genome size results in larger cells. In turn, larger cells result in a simplification of brain morphology. Thus, quite independently of the demands of function, internal features such as genome size can affect the morphology and organization of complex animals. Plethodontid salamanders share the basic vertebrate nervous system and brain, but they have very little space in their small skulls and spinal cords.[31]

The problem is, however, that none of this really excludes a functionalist type of explanation. Raff has to admit that if there are constraints at work, they apparently do not make much difference. The salamanders can do some pretty remarkable things—remarkable salamander things, that is—seeming not at all to be functionally constrained. "These salamanders occupy a variety of caverniculous, aquatic, terrestrial, and arboreal habitats. They possess a full range of sense organs, and most remarkably, a spectacular insect-catching mechanism consisting of a projectile tongue that can reach out in ten milliseconds to half the animal's trunk length (snout to vent is the way herpetologists express it)."

They have pretty good depth perception too. And indeed, their slow metabolic rate brought on by large genome size may even be of adaptive advantage. "Plethodontids are sluggish, and the low metabolic rates introduced by large cell volume may be advantageous to sit-patiently-

and-wait hunters that can afford long fasts. Vision at a distance is reduced to two handbreadths, but since these animals are ambush hunters that strike at short range, that probably doesn't affect their efficiency much."[32] All in all, there is not much for an ardent selectionist to worry about here, and apparently, if need be, the salamanders can start to bring down their genome sizes. The constraints are just not that strong.

DEVELOPMENTAL CONSTRAINTS

This kind of pattern—constraints; need to take seriously formal issues; but not excluding functionalist interpretations—plays itself out over and over again. Consider developmental constraints, and to do this, take the stunning new discoveries in science mentioned at the beginning of this essay. As Darwin pointed out, especially in his little book on orchids, organisms often do not start from scratch. They use what they have at hand and adapt from there. They recycle, in other words. Apparently, at the molecular level, organisms are even greater recyclers of already-available material than they are at the physical level.[33] Most remarkable of all are certain so-called homeotic genes. These are not structural genes—that is, genes that are coded to make the actual bodily products—but developmental genes that are coded to process the production of bodily products by structural genes. The homeotic genes are those that regulate the identity and order of the parts of the body—a mutation in one perhaps moving an eye to where a leg might normally appear, or vice versa. A subclass of such homeotic genes contains "*Hox* genes," a group of genes to be found in bilaterans (organisms the same on both sides), that order the appearance of various bodily parts, and that seem to work in the sequence as they are found on the chromosomes. In *Drosophila*, the *Hox* genes start up at the head, work down through the thorax, and so on to the end of the abdomen. Within these genes, one finds lengths of DNA, of 180 base pairs, that are used to bind the genes to other DNA segments that are part of structural genes. In other words, these "homeoboxes" make a protein (of sixty amino acids)—the "homeodomain"—that is the key to the *Hox* genes actually functioning in regulating the structural genes.

What was absolutely staggering was the discovery of a homology between the homeodomains of *Drosophila* (fruit flies) and other bilaterans, from frogs through fish and mice to humans. Although it has been hundreds of millions of years since humans and fruit flies shared a common ancestor, it is still the case that we use essentially the same chemical mechanisms to order the production of our various bodily parts. The flies' legs and the humans' legs go back to the same processes. The identities are just too great to be chance or even convergence on a common solution. They are homologous—a consequence that leaves some entirely to downplay the significance of selection. The late Stephen Jay Gould, for one, took all of this to be a clear case of the need for a new perspective. Instead of selection, one has certain basic ground plans or archetypes—*Baupläne*—and it is these that really count in evolution, constraining its course and nature. Organisms get certain basic patterns in place, and from then on they cannot escape or do anything outside the constraints set by these unmovable patterns.[34]

FUNCTIONALISTS STRIKE BACK

But is this well taken? Could it not simply be a case of, "If it ain't broke, don't fix it"? In other words, could it not be the case that the way that these genes work is adaptive, and that they are kept in place by selection? It works for fruit flies; it works for humans. End of story. Talk of selection not operating because of "phylogenetic inertia," or some such thing, is simply unwarranted. For all that Rudolf Raff was one of those quoted above about the triumph of the developmental way, it is he who points out that the genetic homologies may in fact not be all that rigid—suggesting that if selection wants to move things functionally, it can. It is not simply a case of being locked in and unable to do anything else.

Raff discusses a certain *Hox* gene (the Antennapedia complex) that occurs in two species of *Drosophila*, *D. melanogaster*, and *D. pseudoobscura*, species that (measured by molecular clocks) parted about 46 million years ago, not that long in the history of life:

> As expected, the clusters of two *Drosophila* species are highly similar in gene composition, arrangement, and conservation of the very long

transcription units characteristic of the *D. melanogaster* Antennapedia complex. However, some revealing differences were found. The *Deformed* locus of *D. pseudoobscura* is inverted with respect to the *Deformed* gene in the *D. melanogaster* cluster. The *D. pseudoobscura* orietation of this gene is the primitive one shared with mammals (all Hox genes transcribed in the same order). It was inverted sometime during the 46 million years that separates the *D. melanogaster* lineage from its congeners.[35]

There are other differences of a like kind, adding up to the conclusion that "the Hox complex can tolerate substantial change over moderate periods of evolutionary time and within a common body plan."

In short, no one will deny that homologies—including molecular homologies—exist and are obviously important. They are a clear sign of evolution and a significant aspect of organic form. But do they—or any like instances of "phylogenetic inertia"—constrain evolution in any significant way? By analogy, could it not be that "the fact that all tires are round more likely means that round wheels are optimally functional than that tire companies are somehow constrained by the round shape of their existing molds. Thus phylogenetic inertia is not an alternative to natural selection as a mechanism of persistence, and evidence of the former is not evidence against the latter"?[36] One needs evidence of more than homology to argue against adaptation. One needs evidence that homology is *preventing* adaptation.

To be frank, Darwinians suspect that there is something of a sleight of hand at work here. The new formalists, people like Stephen Jay Gould, focus on things like the four-limbedness of vertebrates—things that have no obvious adaptive function today. But rather than at once plunging back into a quest for adaptation in the past, when such features first appeared, the formalist focus remains on the present and—thanks to somewhat idiosyncratic (and self-serving) definitions of adaptation that refuse to apply the term to subsequent uses after the initial use—all that occurs is the labeling of four-limbedness today as "nonadaptive." A final move is to think up some fancy name like "archetype" or "*Bauplan*," giving ontological status to what you are promoting, and the dish is complete. Function is relegated to the sidelines.

STRUCTURAL CONSTRAINTS

There are other kinds of putative constraints, including structural constraints—the prime examples of which (in the human world) are the structural constraints of the pillars in medieval churches leading (at the points where pillar meets roof) to areas of no function, the much-discussed (in the opinion of some, the overly much-discussed) "spandrels" of which Gould made so much. Although such spandrels seem adaptive—areas for creative outpourings—in fact they are just by-products of the builders' methods of keeping the roof in place. "The design is so elaborate, harmonious, and purposeful that we are tempted to view it as the starting point of any analysis, as the cause in some sense of the surrounding architecture." This, however, is to put the cart before the horse. "The system begins with an architectural constraint: the necessary four spandrels and their tapering triangular form. They provide a space in which the mosaicist worked; they set the quadripartite symmetry of the dome above."[37] Perhaps, argues Gould, we have a similar situation in the living world. Much that we think adaptive is merely a spandrel, and such things as constraints on development prevent anything like an optimally designed world. Perhaps things are much more random and haphazard—nonfunctional—than the Darwinian thinks possible.

To which the functionalist replies: Who ever thought otherwise? No one since Darwin has ever claimed that everything works just right. It is more the case that one should look first for function, and recognize that sometimes it is more complex than simply one thing or another. After all, things like allometry (where body parts are a logarithmic function of the overall body and hence these parts can grow at a much greater rate than the whole) have long been known and acknowledged, and they are the prime example of trying to put a functioning organism together properly and of having trouble optimizing every last feature. The adult male Irish Elk had a huge set of antlers that may well have been nonadaptive—but this could have come from extreme sexual selection among young males (the Elks were cervine deer, a group where the successful male uniquely keeps and impregnates a whole harem), with much pressure to produce antlers when young, even though when fully adult the antlers proved counterproductive. This would be a case where function worked but then backfired into nonfunctionality.

It has also always been agreed that redundant or unwonted characteristics, produced as part of the process of putting things together, might later be picked up by selection and used in their own right—as of course are the spandrels of San Marco. Pumping testosterone through the bodies of male humans has all sorts of adaptive virtues—penises and testicles for a start. Perhaps it is indeed the case that a secondary effect is hair on the male face, but despite the best advertising efforts of the Gillette razor company, it is not obvious that beards are without their adaptative advantages. Or to use Gould's own favorite example, that the much enlarged clitoris of the female hyena has no present adaptive virtues.[38] It may be that such a pseudopenis came by chance, but this does not deny its value today in mating rituals.

In fact, it has always been a key part of Darwinism that side effects can be a crucial part of the evolutionary story. A big question is about how new characteristics ever get started. As Gould forever asks, could a tenth of an eye be of any great value? Well, as Dawkins responds, perhaps it could, but there is no need to suppose immediate value in every new feature.[39] Feathers today obviously have the adaptive edge when it comes to flying, and no one would say that flight is without its purposes. Increasingly, however, the evidence is that feathers first appeared on the dinosaurs, not for flight, but for other ends, most likely insulation and heat control. Only later were they used to invade the air.

PHYSICAL CONSTRAINTS

There are other potential types of constraint. Particularly interesting are physical constraints. Why do we never get a cat as big as an elephant? Because size and consequent weight goes up rapidly, according to the cube power of length or height. Suppose you have two identically shaped mammals, one twice the height of the other. It is going to be eight times as heavy. This means that from a structural perspective, it has got eight times the weight problem. You simply cannot build elephants as agile as cats. They need far more support, which in turn means bigger and more stocky bones.[40]

Another interesting calculation concerns what has irreverently been called the "Jesus number." What are the constraints on walking on water?

In fact, a fairly simple formula governs the activity. Pushing up is the surface tension, γ, times the perimeter of the feet or area that is touching, l. Pushing down is gravity, which is a function of the mass, m, times the gravitational attraction, g ($F = mg$), or restating in terms of density, ρ, times the volume, which is a function of l cubed. In other words: $Je = \gamma l/\rho l^3 g = \gamma/l^2 g$. Since the surface tension, the density, and the gravitational constant remain the same, this means that the ability to walk on water is essentially a function of the perimeter squared. In other words, the smaller you are the better off you are, and conversely the bigger you are the more likely you are to sink. This is no problem for insects, especially given that they have six legs and so have a lot of perimeter for the body size. Humans however are another matter. "What would be the maximum weight of a human who could walk on water? My size 9 sandals have a perimeter of 0.62 meters each; that length times the surface tension of water gives 0.045 newtons of force, or 4.6 grams (less than half an ounce) of weight—9.2 grams to stand (two feet in contact) or half that to walk. The theological implications are beyond the scope of the present book."[41] This essay too!

However, the philosophical implications are clear. There is more to life than simple function, but even in a case like this the functionalist will see selection at work. The point is that organisms exploit physical constraints for their own ends. John Maynard Smith and his coauthors have explored in depth the example of the coiling of shells in such organisms as mollusks and brachiopods. The coiling itself is fairly readily reduced to a simple logarithmic equation, and it is possible to draw a plan that maps the coiling as a function of the vital causal factors, particularly the rate of coiling and the size of the generating curve. Given such a map, one feature stands right out for comment: whereas for most shells the coils touch all the way from the center to the perimeter, some such shells coil without touching. There is a gap between the coils. Now, map the actual shells of a group of organisms, for instance, the genera of extinct ammonoids (cephalopod mollusks), including the theoretical line dividing touching coils from nontouching coils. The isomorphism between the theoretical and the actual is outstanding:

> [N]early all ammonoids fall on the left side of the [theoretical] curved
> line and thus display overlap between successive whorls. This is clearly
> a constraint in the evolution of the group but what kind of constraint?

> In this particular case, the answer is apparently straightforward. . . .
> Evolving lineages can and occasionally do cross the line so there is no
> reason to believe that open coiling violates any strict genetic or devel-
> opmental constraint. Rather, the reason for not crossing the line
> appears to be biomechanical. Other things being equal, an open coiled
> shell is much weaker than its involute counterpart. Also, open coiling
> requires more shell material because the animal cannot use the outer
> surface of the previous whorl as the inner surface of the new whorl.[42]

Even the exceptions prove the point. The shell of the living pelagic
cephalopod Spirula has a shell that coils but does not touch. Exceptionally, this organism carries the shell internally, using it for buoyancy.
There is no need for strength.

Maynard Smith and coauthors conclude "that the constraint against
open coiling is an adaptive one brought about by simple directional
selection." A conclusion that surely brings us full circle, for if constraints
can be *adaptive*, brought on by selection, the distinction between form
and function has truly collapsed. Indeed, theoretical biologist Gunter
Wagner goes so far as to argue that constraints may be necessary for the
action of selection, else variation will be all over the place with any positive changes being outweighed by other moves in a wrong direction.[43]

CONCLUSION

Enough has been said. Around about 1959, the centenary of the publication of the *Origin*, functionalism was just about the only game in town.
Apart from a few despised German morphologists and one or two
idiosyncratic philosophers, no one had any time for a formalist approach
to nature. Now things have swung the other way, and in some evolutionary circles, formalism rules triumphant. However, pulling back and
looking at biology as an outsider—as a philosopher, that is—one can see
that things are not quite this simple. As the above discussion shows, a
case can be made for formalism. But equally, as the above discussion
shows, a case can still be made for functionalism.

In this sense, one should probably not speak of rival paradigms, but
rather of different elements that go to make up the overall complete pic-

ture. But one senses that, historically and philosophically, there is more to be said than this. Historically, the form-function dichotomy—form-function *tension*, rather—goes back to Aristotle and forward to today, right across the coming of evolution. And philosophically one does see some of the features of a paradigm, inasmuch as—rather like the Gilbert and Sullivan operetta (*Iolanthe*), where "every boy and every gal, that's born into the world alive, is either a little Liberal, or else a little Conservative"—biologists do seem to split over whether one should think form first and function second (and reluctantly at that), or function first and form second (and reluctantly at that).

Darwin, to take one example, was an out-and-out functionalist. He recognized form—he spoke of homology as Unity of Type—but saw it as coming from function. Gould, to take the other extreme, was an out-and-out formalist, and he saw function as coming a distant second. His hero was the early twentieth-century Scottish morphologist, D'Arcy Thompson, who had no sympathy with the Darwinian's focus on final cause and who argued strenuously that material and formal causes are prior and more important:

> To seek not for ends but for antecedents is the way of the physicist, who finds "causes" in what he has learned to recognise as fundamental properties, or inseparable concomitants, or unchanging laws, of matter and of energy. In Aristotle's parable, the house is there that men may live in it; but it is also there because the builders have laid one stone upon another.[44]

Continuing:

> Cell and tissue, shell and bone, leaf and flower, are so many portions of matter, and it is in obedience to the laws of physics that their particles have been moved, moulded and conformed. . . . Their problems of form are in the first instance mathematical problems, their problems of growth are essentially physical problems, and the morphologist is, *ipso facto*, a student of physical science.[45]

Thus: "We want to see how, in some cases at least, the forms of living things, and of the parts of living things, can be explained by physical considerations, and to realise that in general no organic forms exist save such as are in conformity with physical and mathematical laws."[46]

We have different perspectives, different visions. In this sense, we do have different paradigms. And, let me stress, in case you think I am ambiguous or disapproving, this is no bad thing. I do not see this as a sign of weakness in biology or of immaturity. If a division has been around for 2,500 years, and is thriving today as never before, I doubt it is going to go away in a hurry. But the point I would make is that because biologists are motivated to go out and make their case—form or function—some terrific science gets done. The division is incredibly creative. So let us not regret it, but rejoice in it. The task now for philosophers is to try to understand it. What is this modified notion of "paradigm" with which I am playing? Is it unique to biology or can it be found elsewhere in science? And what does this tell us about such issues as objectivity and subjectivity? These are the questions to which we philosophers should turn now.[47]

NOTES

1. S. B. Carroll, J. K. Grenier, and S. D. Weatherbee, *From DNA to Diversity: Molecular Genetics and the Evolution of Animal Design* (Oxford: Blackwell, 2001).

2. E. S. Russell, *Form and Function: A Contribution to the History of Animal Morphology* (London: John Murray, 1916; repr. University of Chicago Press, 1982); M. Ruse, "Is the Theory of Punctuated Equilibria a New Paradigm?" *Journal of Social and Biological Structures* 12 (1989): 195–212.

3. Aristotle, *Physics* 194b 30–195a 1, in J. Barnes, ed., *The Complete Works of Aristotle* (Princeton, NJ: Princeton University Press, 1984), 1: 323–33.

4. Aristotle, *Parts of Animals*, 641a 7–17, in ibid., 1:997.

5. Aristotle, *History of Animals*, 498b 11–13, in ibid., 1:793.

6. J. Ray, *Wisdom of God, Manifested in the Words of Creation*, 5th ed. (London: Samuel Smith, 1709 [1691]).

7. Ibid., pp. 32–33.

8. Ibid., pp. 335–36.

9. Letter to his publisher, John Murray, September 24, 1861, in C. Darwin, *The Correspondence of Charles Darwin* (Cambridge: Cambridge University Press, 1985), 9:279.

10. C. Darwin, *On the Various Contrivances by which British and Foreign Orchids Are Fertilized by Insects, and On the Good Effects of Intercrossing* (London: John Murray, 1862), p. 15.

11. Ibid., p. 16.

12. Ibid., p. 65.

13. Ibid., p. 182.

14. C. Darwin, *On the Origin of Species by Means of Natural Selection, or the Preservation of Favoured Races in the Struggle for Life* (London: John Murray, 1859), p. 63.

15. Ibid., pp. 80–81.

16. R. J. Richards, *The Romantic Conception of Life: Science and Philosophy in the Age of Goethe* (Chicago: University of Chicago Press, 2003); T. A. Appel, *The Cuvier-Geoffery Debate: French Biology in the Decades Before Darwin* (New York: Oxford University Press, 1987); M. Ruse, *Monad to Man: The Concept of Progress in Evolutionary Biology* (Cambridge, MA: Harvard University Press, 1996).

17. Rev. R. Owen, *The Life of Richard Owen* (London: Murray, 1894), 1:388).

18. T. H. Huxley and H. N. Martin, *A Course of Practical Instruction in Elementary Biology* (London: Macmillan, 1875).

19. F. M. Balfour, *A Treatise on Comparative Embryology* (London: Macmillan, 1880–1881), 2: 214.

20. Ibid., p. 177.

21. C. Gegenbaur, *Elements of Comparative Anatomy* (London: Macmillan, 1878), p. 4.

22. E. C. Agassiz, ed., *Louis Agassiz: His Life and Correspondence* (Boston: Houghton Mifflin, 1885), 1:241.

23. Ruse, *Monad to Man.*

24. R. J. Richards, *The Meaning of Evolution: The Morphological Construction and Ideological Reconstruction of Darwin's Theory* (Chicago: University of Chicago Press, 1992).

25. Agassiz, *Louis Agassiz,* 1:369–70.

26. Ruse, *Monad to Man*; M. Ruse, *Mystery of Mysteries: Is Evolution a Social Construction?* (Cambridge, MA: Harvard University Press, 1999).

27. R. Dawkins, "Richard Dawkins: A Survival Machine," in *The Third Culture*, ed. J Brockman (New York: Simon and Schuster, 1983), p. 404.

28. Ibid.

29. J. Maynard Smith et al., "Developmental Constraints and Evolution." *Quarterly Review of Biology* 60 (1985): 269.

30. R. Raff, *The Shape of Life: Genes, Development, and the Evolution of Animal Form* (Chicago: University of Chicago Press, 1996), p. 304.

31. Ibid., p. 305, referring to G. Roth, J. Blanke, and D. B. Wake, "Cell Size Predicts Morphological Complexity in the Brains of Frogs and Salamanders," *Proceedings of the National Academy of the Sciences, USA* 91 (1994): 4796–4800.

32. Raff, *Shape of Life*, p. 306.

33. S. B. Carroll, "Homeotic Genes and the Evolution of Arthropods," *Nature* 376 (1995): 479–85; Carroll, Grenier, and Weatherbee, *From DNA to Diversity*.

34. S. J. Gould, *The Structure of Evolutionary Theory* (Cambridge, MA: Harvard University Press, 2002).

35. Raff, *Shape of Life*, p. 309.

36. H. K. Reeve and P. W. Sherman, "Adaptation and the Goals of Evolutionary Research," *Quarterly Review of Biology* 68 (1993): 18.

37. S. J. Gould and R. C. Lewontin, "The Spandrels of San Marco and the Panglossian Paradigm: A Critique of the Adaptationist Programme," *Proceedings of the Royal Society of London, Series B: Biological Sciences* 205: 582.

38. Gould, *Structure of Evolutionary Theory*.

39. R. Dawkins, *The Blind Watchmaker* (New York: Norton, 1986).

40. S. Vogel, *Life's Devices: The Physical World of Animals and Plants* (Princeton, NJ: Princeton University Press, 1988).

41. Ibid., p. 100.

42. Maynard Smith et al., "Developmental Constraints and Evolution," p. 280.

43. G. P. Wagner, "The Influence of Variation and of Developmental Constraints on the Rate of Multivariate Phenotypic Evolution," *Journal of Evolutionary Biology* 1 (1988): 45–66.

44. D. W. Thompson, *On Growth and Form*, 2nd ed. (Cambridge: Cambridge University Press, 1948), p. 6.

45. Ibid., p. 10.

46. Ibid., p. 15.

47. I do not pretend that I am the first to discover the kind of entity that I am calling a "paradigm." In a way, it seems to me to be much like what Karl Popper called a "metaphysical research programme." ("Darwinism as a Metaphysical Research Programme," in *The Philosophy of Karl Popper*, vol. 1, ed. P. A. Schilpp [LaSalle, IL: Open Court], pp. 133–43.) I am not, of course, claiming that Popper would agree that form and function are exemplars of such a notion. One thing that does seem important to this notion as I am using it is the idea of metaphor. We are seeing the world as if it were an object of design—utilitarian or formal. I explore this kind of thought in my *Mystery of Mysteries: Is Evolution a Social Construction?* (Cambridge, MA: Harvard University Press, 1999). I note with some interest that Kuhn himself tied in paradigm thinking with metaphor ("Metaphor in Science," in *Metaphor and Thought*. 2nd ed., ed. Andrew Ortony [Cambridge: Cambridge University Press, 1993], pp. 533–42).

Nine

DARWINISM EXPLAINS RELIGION (?)

arwinian evolutionary theory has always taken behavior seriously. From the beginning, and especially in the *Origin*, Darwin realized that what animals do is as important as what they are. Being fleet of foot can be as advantageous in the struggle for existence and the ability to digest (say) grass or meat. Darwinian evolutionary theory has always applied to humans, so human behavior has always been a topic of interest. Note therefore the crucial causal plank, the key method of inquiry. For the Darwinian, natural selection is the first-row key to everything. It may not always work but it is the first thing to be invoked. And, the important thing about selection is that it does not merely lead to change but to change of a particular kind, namely, in the direction of adaptation or contrivance. Selection produces things that function toward desired ends, such as the eye being used for seeing and the teeth being used for biting and chewing. Hence, for the Darwinian interested in human behavior, the key to understanding is adaptation, brought on by natural selection. However, the key is not all-powerful. Complicating the picture is the fact that not all features of the living world are necessarily adaptive. Some occur by chance and some are by-products of selection. So a major part of the Darwinian's task is determining if something is adaptive and hence probably produced by selection, or if something is not adaptive and in which case what did cause it, if indeed there was an identifiable cause. This obviously applies very much to studies of human behavior.

Religion is a major factor in human behavior and culture and naturally it has attracted considerable Darwinian attention. The big problem, therefore, is whether or not it is adaptive and if so, in what way, and if not, why then does it exist. As always, it is best to start with Charles Darwin himself to set the background.

DARWIN ON RELIGION

Much ink has been spilt on the question of Charles Darwin's debt to David Hume. My own feeling is that, although clearly Hume's general empiricism was important, and although many things in Hume cry out for evolutionary understanding (I do not think that Hume was an evolutionist), we should not overemphasize the connections. There were other more immediate philosophical influences—John F. W. Herschel and William Whewell, for instance. Sir James Mackintosh in ethics, perhaps. (For more details, look at Darwin's *Autobiography*.)[1] However, one place where Hume did have a major influence was on Darwin's thinking about the natural origins of religion. We know that, as a young man, Darwin read Hume's *Natural History of Religion*, and when Darwin himself took up the problem of its origins, the Scottish skeptic's heavy accent can almost be heard up from the pages.

It is in his *Descent of Man* that Darwin turns to the problem of religion—not as something to be justified or attacked, but as something to be explained. Thinking that the most primitive form of religion is when savages believe in spirit forces, he asks about its origins. Apparently it is all a question of seeing spirits in inanimate objects, feeling or pretending or mistakenly believing that they are truly alive:

> The tendency in savages to imagine that natural objects and agencies are animated by spiritual or living essences, is perhaps illustrated by a little fact which I once noticed: my dog, a full-grown and very sensible animal, was lying on the lawn during a hot and still day; but at a little distance a slight breeze occasionally moved an open parasol, which would have been wholly disregarded by the dog, had any one stood near it. As it was, every time that the parasol slightly moved, the dog growled fiercely and barked. He must, I think, have reasoned to himself in a

rapid and unconscious manner, that movement without any apparent cause indicated the presence of some strange living agent, and that no stranger had a right to be on his territory.[2]

Darwin did not think that explanations of religious belief bore on the truth or falsity of religion. Whether true or false, the important point is that religion as considered by the scientist be considered a natural phenomenon. He stressed this point again and again:

> I am aware that the assumed instinctive belief in God has been used by many persons as an argument for His existence. But this is a rash argument, as we should thus be compelled to believe in the existence of many cruel and malignant spirits, only a little more powerful than man; for the belief in them is far more general than in a beneficent Deity. The idea of a universal and beneficent Creator does not seem to arise in the mind of man, until he has been elevated by long-continued culture.[3]

Darwin's position, therefore, was that religion is a natural phenomenon—or rather, a phenomenon that can be treated naturally—and he saw it as something that had evolved. It is noteworthy that Darwin said little about religion and its relationship to natural selection. Here there is a major break from Darwin's parallel discussion of morality, which did get linked firmly to selection. Perhaps for all that he claimed not to be addressing the truth status of religion, because by this time in his life Darwin had become an agnostic—certainly he thought that Christianity is not proven in its essentials and false in many details—he did not think that religion could be directly promoted by selection. For Darwin, religion seems to be almost accidental, and brought about by animal features or powers that are simply misdirected. When we see something moving, it normally makes sense to think that it is living. We make mistakes, and ultimately this leads into religion. The one concession that he was prepared to make is that, in the case of civilized people, religion does help reinforce morality: "With the more civilised races, the conviction of the existence of an all-seeing Deity has had a potent influence on the advance of morality."[4] But even here, Darwin did not want to explore the matter in more detail.

RELIGION AS ADAPTIVE

In the hundred years after the *Origin*, there was much interest in putative natural origins of religion. But most discussion came from the newly developing social sciences rather than from biology. The growth of Darwinian studies of social behavior—what is often known as "sociobiology"—has changed all of that. As always in discussions of evolution it is natural selection that drives the thread of the investigation. Hence, let us be guided by biological categories—Darwinian categories, that is. Most importantly, we will expect a division between those who think that religion is something brought about directly by natural selection, and those (like Darwin) who think that religion is something of a by-product. Then among those who suppose selection as the cause, there will be division between those who think that religion is of direct adaptive advantage to humans and those who think that it might not be such a good thing to have—perhaps a product of something like sexual selection or perhaps adaptive for someone or thing other than humans (like parasites). There is also the possibility of division between individual and group selectionists—that is, between those who think that religion must be for the benefit of the individual and those who think that religion is of group worth, perhaps even to the detriment of the individual. Finally, there is possibility of division between those who think that religion is essentially biological and those who think that culture is significant if not all-important.

Starting with those who think that religion is selection produced and of value to humans, we find the grand old man of Darwinian social studies, of sociobiology, Edward O. Wilson. For him, religion is apparently all a matter of group identity and sticking together:

> The highest form of religious practice, when examined more closely, can be seen to confer biological advantage. Above all, they congeal identity. In the midst of the chaotic and potentially disorienting experiences each person undergoes daily, religion classifies him, provides him with unquestioned membership in a group claiming great powers, and by this means gives him a driving purpose in life compatible with his self interest.[5]

Wilson does allow that there can be cultural selection between sects, but essentially we start with the biology and all else is on the surface:

Because religious practices are remote from the genes during the development of individual human beings, they may vary widely during cultural development. It is even possible for groups, such as the Shakers, to adopt conventions that reduce genetic fitness for as long as one or a few generations. But over many generations, the underlying genes will pay for their permissiveness by declining in the population as a whole. Other genes governing mechanisms that resist decline of fitness produced by cultural evolution will prevail, and the deviant practices will disappear. Thus culture relentlessly tests the controlling genes, but the most it can do is replace one set of genes with another.[6]

Wilson has always been ambivalent about the comparative significances of individual- and group-selective processes. On the religion issue, he rather divides, thinking that it is something brought on by a group process, but surely with individual benefits also. More robustly individualistic are the physical anthropologist Vernon Reynolds and the scholar of religion Ralph Tanner.[7] They are quite accepting of such hypotheses as that circumcision of males, a practice central to religion of Jews and others, is something that prevents disease. This is a practice that benefits individuals. Somewhat ingeniously, Reynolds and Tanner suggest that religions tend to divide into those that promote high reproductive rates—many Semitic religions—and those that do not—North-European Calvinism, for instance. This is something echoing interests and concerns of Darwin in the *Descent*. There, the great evolutionist worried that the worthless Catholic Irish seem to have lots of children whereas the hardworking Presbyterian Scots have but few. This was a horrific reflection, seemingly negating the upward, progressive nature of the evolutionary process, a picture so dear to the heart of Darwin and his fellow Victorians. Darwin consoled himself with the reflection that the Irish do not look after their kids whereas the Scots do, and so on balance the Scots if anything do better than the Irish.

Reynolds and Tanner pick up on this insight, drawing on its modern equivalent, the so-called r and K selection theory. The herring is an r-selectionist, having many offspring but not caring for them. The elephant is a K selectionist, having but few offspring and investing a great deal of parental care. As part of this theory, today's evolutionists suggest that if conditions are highly variable then a good reproductive strategy gets one to have lots of offspring albeit with little individual attention (r-selection), whereas if conditions are stable then the better strategy is few children and much care (K-selection). Variable conditions mean that sometimes you might strike it rich; whereas with few offspring you might never do that well—and conversely. In the case of male circumcision, acknowledging that there are questions about the evidence, Reynolds and Tanner nevertheless write:

> Despite the confused state of the data, it is not unreasonable to put the question: If circumcision does reduce the risk of penile or cervical carcinoma, what effects would this have on reproductive success? The answer is that such success should be increased (all other things being equal) in families or groups practising circumcision. . . . In the case of Judaism it represents part of Abraham's Covenant with God, the covenant in which God called him to leave Ur and to found a new nation; also in the Covenant was the promise from God that his 'seed' would inherit the land.[8]

The idea seems to be that by moving to Israel, Abraham was moving to a land where the chances of raising children moved from very variable to more certain (albeit within restricted limits), and hence there was a move from r-selection-type practices to K-selection-type practices, one of the latter being circumcision.

GROUPS AND MEMES AS UNITS OF SELECTION

Showing just how different people's thinking can be and yet still (in the eyes of advocates) be under the banner of Darwinism—that is, people who would think of themselves as working in the sociobiological mode—we have the biologist David Sloan Wilson and the philosopher Daniel Dennett. Wilson is openly committed to a group-selective analysis of religion, wanting to regard societies as akin to organisms and as strengthened by a sincere commitment to a religious doctrine. He ties this thesis strongly to morality, which he speaks of as having "both a genetically evolved component and an open-ended cultural component."[9] Wilson analyzes the society that Jean Calvin founded in Geneva in the sixteenth century, listing the rules that governed this group: "Obey parents," "Obey magistrates," "Obey pastors," and on down the list to "No lewd-ness and sex only in marriage," "No theft, either by violence or cunning," and so forth. Of this Wilson writes:

> To summarize, the God-people relationship can be interpreted as a belief system that is designed to motivate the behaviors [examples of which are listed just above]. Those who regard religious belief as sense-less superstition may need to revise their own beliefs. Those who regard supernatural agents as imaginary providers of imaginary services may have under-estimated the functionality of the God-person relationship in generating real services that can be achieved only by communal effort. Those who already think about religion in functional terms may be on the right track, but they may have underestimated the sophistication of the "motivational physiology" that goes far beyond the use of kinship terms and fear of hell. Indeed, it is hard for me to imagine a belief system better designed to motivate group-adaptive behavior for those who accept it as true. When it comes to turning a group into a societal organism, scarcely a word of Calvin's catechism is out of place.[10]

Although he thinks that culture is crucial, ultimately Wilson sees real change as genetic. Coming in a very different direction, the philosopher Daniel Dennett agrees entirely that that religion is something promoted by selection, but he is not at all convinced that this selection is necessarily for the benefit of humans, nor is it essentially (or truly in any way) genetic. Dennett has adopted a theory of Richard Dawkins that posits the existence of "memes," units of culture akin to genes, which compete for people's allegiances.[11] Rival memes, as it were, invade people's minds and those that win are those that are selected to continue. Winning is not random but a function of the features—the adaptations—that the memes have or promote. Successful advertising obviously is a paradigmatic example of memes at work—one buys and smokes Marlborough cigarettes because this makes you feel like a real man, even though in fact you are acting in ways highly detrimental to your health and well-being.

For Dennett, religion is a meme parasite that has features that make it attractive even if it is not necessarily that good for the possessor. So that no one miss this point, he begins his book *Breaking the Spell: Religion as a Natural Phenomenon* by introducing the reader to the lancet fluke

(*Dicrocelium dendriticum*), a parasite that corrupts the brain of an ant, causing it to strive to climb blades of grass, that it get eaten by a sheep or cow, and thus the fluke can complete its life cycle before its offspring are excreted and take up again with ants:

> Does anything like this ever happen with human beings? Yes indeed. We often find human beings setting aside their personal interests, their health, their chances to have children, and devoting their entire lives to furthering the interests of an *idea* that has lodged in their brains. The Arabic word *islam* means "submission," and every good Muslim bears witness, prays five times a day, gives alms, fasts during Ramadan, and tries to make the pilgrimage, or *hajj*, to Mecca, all on behalf of the idea of Allah, and Muhammad, the messenger of Allah. Christians and Jews do likewise, of course, devoting their lives to spreading the Word, making huge sacrifices, suffering bravely, risking their lives for an idea. So do Hindus and Buddhists.[12]

To be fair, Dennett adds that secular humanists are often not much better in this regard.

RELIGION AS BY-PRODUCT

What of those who think religion falls more into the by-product category? The late Stephen Jay Gould himself was one who thought along these lines. The whole of human culture came under this category for him.[13] But most would not be this sweeping. Apart from anything else, religion with its costs—devotion to others, celibacy, ritual physical disfigurement, sacrifice, and so forth—simply does not seem to be the sort of thing that just happened as a by-product. It is just too costly. More likely is the idea that religion, as it were, piggybacks into existence and power on the backs of other things—real, powerful adaptations—and manages to exist because it cannot be stopped or because ultimately its costs are simply not that great. Student of culture Pascal Boyer inclines to the first option. For him, religion simply subverts or borrows features that our biology has put in place for good adaptive reasons, and for whatever reason it cannot be eradicated:

The building of religious concepts requires mental systems and capacities that are there anyway, religious concepts or not. Religious morality uses moral intuitions, religious notions of supernatural agents recruit our intuitions about agency in general, and so on. This is why I said that religious concepts are parasitic upon other mental capacities. Our capacities to play music, paint pictures or even make sense of printed ink-patterns on a page are also parasitic in this sense. This means that we can explain how people play music, paint pictures and learn to read by examining how mental capacities are recruited by these activities. The same goes for religion. Because the concepts require all sorts of specific human capacities (an intuitive psychology, a tendency to attend to some counterintuitive concepts, as well as various social mind adaptations), we can explain religion by describing how these various mind capacities get recruited, how they contribute to the features of religion we find in so many different cultures. We do not need to assume that there is a *special* way of functioning that occurs only when processing religious thoughts.[14]

But what is it that allows religion to get its hold in the first place? Anthropologist Scott Atran inclines to the second option, that religion grabs something adaptively useful and exploits it. For him, the big question facing organisms like humans is other living beings—above all, other living beings as threats. In an argument reminiscent of Darwin and his dog, Atran suggests that what we have is a somewhat overeager projection of the living onto the inanimate. It used to be thought that the baroque nasal appendages of the titanotheres were a case of sensible evolution having taken a step too far. Perhaps the same is true of religion. Cuckoos exploit the innate mechanisms that their host birds have for raising their young. Religion does much the same for humans:

> Supernatural agent concepts critically involve minimal triggering of evolved agency-detection schema, a part of folk psychology. Agency is a complex sort of "innate releasing mechanism." Natural selection designs the agency-detection system to deal rapidly and economically with stimulus situations involving people and animals as predators, protectors, and prey. This resulted in the system's being trip-wired to respond to fragmentary information under conditions of uncertainty, inciting perception of figures in the clouds, voices in the wind, lurking movements in the leaves, and emotions among interacting dots on a

computer screen. This hair-triggering of the agency-detection mechanism readily lends itself to supernatural interpretation of uncertain or anxiety-provoking events.

People interactively manipulate this universal cognitive susceptibility so as to scare or soothe themselves and others for varied ends. They do so consciously or unconsciously and in causally complex and distributed ways, in pursuit of war or love, to thwart calamity or renew serendipity, or to otherwise control or incite imagination. The result provides a united and ordered sense for cosmic, cultural, and personal existence.[15]

SERIOUS SCIENCE?

What can we say about these various ideas and hypotheses? One thing for certain: they can't all be true! For every action there is an equal and opposite reaction. For every idea about the evolution of religion, there is an idea that takes exactly the opposite tack! Science can break down for two basic reasons: the theories are no good, or the evidence is not supportive. Both of these reasons come into play with the sociobiological accounts of religion.

With respect to theory, straight Darwinism goes from strength to strength. The same cannot always be said of the ideas used to explain religion. Take the theory of memes. It really is crude to the point of nonbeing, certainly to the point of the nonhelpful. What is a meme? It is a chunk of culture analogous to a gene. As it happens, genes are hard enough to define, but we do have some idea of them as the smallest functioning length of DNA. But what is the smallest functioning length of culture? Is Catholicism a meme? Is the authority of the pope a meme? Is transubstantiation a meme? Why the authority of the pope, for example, rather than each and every one of the dogmas that he endorses? And what kind of theory do you have as memes clash and come together and sometimes fuse and sometimes break apart? How is Mormonism a meme as compared to evangelical Christianity? Does Mormonism somehow include a lot of the evangelical Christianity meme, or are they separate memes? And so on and so forth. The point is not that Dennett is necessarily wrong in arguing that ideas sometimes have lives of their own, or that religions can be dreadful things that take over people's minds to people's own detriment—if not Catholicism, then most of us think this way about cults like Scientology—but that memetics is not very helpful in understanding what is going on. One is really just taking regular language and putting it in fancy terms. No new insights. No new predictions. No astounding claims that turn out to be true.

There are analogous weaknesses with evidence. One might praise David Sloan Wilson for wanting to turn to real examples to articulate and flesh out his thinking. But his discussion of Calvinism really will not do. If Calvinism was such a terrific booster of societies and helped them work so well, why did it so frequently fail to convince? Take the English: Henry the Eighth broke from the Catholic Church because he wanted to take a new wife and the pope forbade it. His son Edward the Sixth was ultra-Protestant, more Lutheran than anything else. But when Edward died as a teenager, his older sister Mary came to the throne and, as an ardent Catholic, persecuted Protestants, many of whom fled to the continent. By that time, the middle of the sixteenth century, the German Lutheran areas were often violent and torn by war and strife, and so these exiles headed for safer, quieter, Reformed (Calvinist) areas. When Mary died and her younger, Protestant sister Elizabeth came to the throne, the Calvinists all flooded back. But generally the English were not that keen on what they had to offer. They did not want the repressive morality and lifestyle of those who later came to be known as Puritans. So we had the Elizabethan compromise, an Anglican Church—to this day a bricolage of Catholic style and Protestant theology. And it was certainly not that unsuccessful. The English saw off the Spanish and their armada. It is true that in the seventeenth century the Roundheads, the Puritans, won the Civil War and lopped off the head of King Charles the First, but within twelve years the Royalists, the more central Anglicans, were restored and the Puritans were out again.[16]

More generally, Darwinian would-be explainers of religion are far too given to isolating one bit of history, one place in time and space, and thinking you have the basis for a whole theory. This comment applies particularly to Americans—Edward O. Wilson and Daniel Dennett come at once to mind—who start with assumptions about the universal appeal and force of religion. By any measure, given its anti-Enlightenment obsession with religion, America is a very peculiar country, at least compared to the rest of the First World. It is very dangerous to argue about the need that humans have of religion if in fact a lot of humans really do not seem to need that much religion at all. England is a case in point. Most young couples want a church wedding. But that is about the limit to their involvement with religion until the time comes to travel to the crematorium. Generally, religion to the lives of the average person in England is about as relevant as the royal family. Significantly, the queen and her family are expected to observe the ritual practices as a proxy for the rest of us. Not all writers on the biology of religion are totally provincial in thinking that their home society is the norm—not even that all American writers on the biology of religion are equally provincial in thinking that their home society is the norm—but it is an issue inadequately addressed.

GOD?

Suppose that there is something to the naturalistic Darwinian approach to religion, its history, and its nature. How does this cash out philosophically? What does this tell us about God, His nature, and His existence? You might flip the argument entirely on its head, showing on non-Darwinian grounds that God does not exist and then setting forth on a naturalistic journey to explain why nevertheless so many people persist in believing that He is real. This is the tack taken by Dennett. He trots through the various arguments for the existence of God, following through with the standard objections. Then, God dismissed, Dennett is ready to give an argument about why we are deceived. It is hardly surprising that for Dennett religion is a parasite like the lancet fluke.

What if you do not want to go down that path? Ask the basic question about God and His existence. Not the question about whether God exists, but whether a Darwinian account of origins shows that God exists or not. Was Darwin right in thinking that the reality or not of God is irrelevant to a naturalistic account of religion? Edward O. Wilson goes entirely the other way: Darwinism gives a naturalistic account of religion and that is an end to religion as a true description. As it happens, Wilson thinks that the human psyche demands religion, and thus he sees the place to move in with a kind of evolutionary humanism. But this is because Darwinism has already done its corrosive work:

> But make no mistake about the power of scientific materialism. It presents the human mind with an alternative mythology that until now has always, point for point in zones of conflict, defeated traditional religion. Its narrative form is the epic: the evolution of the universe from the big bang of fifteen billion years ago through the origin of the elements and celestial bodies to the beginnings of life on earth. The evolutionary epic is mythology in the sense that the laws it adduces here and now are believed but can never be definitively proved to form a cause-and-effect continuum from physics to the social sciences, from this world to all other worlds in the visible universe, and backward through time to the beginning of the universe. Every part of existence is considered to be obedient to physical laws requiring no external control. The scientist's devotion to parsimony in explanation excludes the divine spirit and other extraneous agents. Most importantly, we have come to the crucial stage in the history of biology when religion itself is subject to the explanations of the natural sciences. As I have tried to show, sociobiology can account for the very origin of mythology by the principle of natural selection acting on the genetically evolving material structure of the human brain.

> If this interpretation is correct, the final decisive edge enjoyed by scientific naturalism will come from its capacity to explain traditional religion, its chief competition, as a wholly material phenomenon. Theology is not likely to survive as an independent intellectual discipline.[17]

One suspects that Wilson is wrong and Darwin was right. The fact that you can give a naturalistic explanation of religion does not at once imply that religion is false. I can give a naturalistic explanation of my belief that the truck is bearing down on me, but it does not follow that the truck is not bearing down on me. It is true that if all you have is a naturalistic explanation, then (Dennett-like) you will probably not be eager to embrace religion. If you can show that religion is indeed a parasite on the mind, why take it any more seriously than the hucksters' e-mail claims that their potions will increase your penis length? But for the traditional religious person—at least, for the traditional Christian religious person—religion has another source of epistemic power that e-mail spam does not have: faith. This being so, then far from a naturalistic account being threatening, many expect such a naturalistic explanation of origins. God had to impart the information to humankind in some way, and why not through evolution? Nor would it be a counterargument that the explanation might make the arrival of religion rather less than edifying— like a dog barking at a parasol blowing in the wind. The job is done.

What about something that often comes up in naturalistic discussions of the origins of religion, namely, the comparative issue? The Christians believe one thing, the Jews another, and the Muslims a third. Now, you might think that this is a pretty good argument against all of them. How can the Christian God be so loving and insist that we acknowledge and worship him, condemning to eternal damnation all of those Asians who grew up in ignorance? But even if you do accept the argument against God based on comparative religion, note that this has nothing whatsoever to do with evolution. It was an argument that moved the deists at the end of the seventeenth century. Moreover, evolution or not, the believer can continue to believe in the face of religious diversity—the Christians (or whatever) got it right and the others did not, and that is the end of matters.

CONCLUSION

If evolution is true, and it is, and if natural selection is the main mechanism, and it is, then the Darwinian approach to religion cannot be without merit. But it has far to go before it can command assent and respect.

NOTES

1. C. Darwin, *Autobiography* (New York: Norton, 1969).
2. C. Darwin, *The Descent of Man, and Selection in Relation to Sex* (London: John Murray, 1871), 1:67.
3. Ibid., 2:394–95.
4. Ibid., 2:394.

5. E. O. Wilson, *On Human Nature* (Cambridge, MA: Harvard University Press, 1978), p. 188.

6. Ibid., p. 178.

7. V. Reynolds and R. Tanner, *The Biology of Religion* (London: Longman, 1983).

8. Ibid., p. 240.

9. D. S. Wilson, *Darwin's Cathedral* (Chicago: University of Chicago Press, 2002), p. 119.

10. Ibid., p. 105.

11. R. Dawkins, *The Selfish Gene* (Oxford: Oxford University Press, 1976).

12. D. C. Dennett, *Breaking the Spell: Religion as a Natural Phenomenon* (New York: Viking, 2006), p. 4.

13. S. J. Gould, *The Structure of Evolutionary Theory* (Cambridge, MA: Harvard University Press, 2002).

14. P. Boyer, *Religion Explained: The Evolutionary Origins of Religious Thought* (New York: Basic Books, 2002), p. 311.

15. S. Atran, *In Gods We Trust: The Evolutionary Landscape of Religion* (New York: Oxford University Press, 2004), p. 78.

16. D. MacCulloch, *The Reformation: A History* (New York: Viking, 2004).

17. Wilson, *On Human Nature*, p. 192.

EVOLUTION AS RELIGION

Are The Creationists Right?

Perhaps there should be a First Rule of Science Journalism:
"Interview at least one person other than Michael Ruse."
—Richard Dawkins, *The God Delusion*

iven the time that I have spent in the past thirty years fighting American Fundamentalism in its various guises—lectures, debates, court appearances, op-ed pieces, essays, books—you would think I might be a hero of the evolutionists. Not a bit of it! There is one group, one very vocal group, who regard me with something midway between pity and scorn, who think that what I am doing is worse than inaction, is something that deserves nothing but condemnation and unremitting opposition. In his smash-hit bestseller *The God Delusion*, Dawkins graciously allows that people like me are not "necessarily dishonest," but he likens me to Neville Chamberlain, the pre–World War II British prime minister who tried to appease Adolf Hitler.[1] Philosopher Daniel Dennett thinks I am taken in by Creationism no less than Brer Fox when Brer Rabbit begged not to be tossed onto the briar patch.[2] Leading University of Chicago evolutionist Jerry Coyne (whose work I used to argue that Sewall Wright's Shifting Balance Theory has no connection with the real world) reviewed one of my books saying: "'One has to belong to the intelligentsia to believe things like that,' George Orwell wrote (in a quite different context). 'No ordinary man could be such a fool.'"[3]

215

This is all a little odd. It really is. The late Henry Morris, leading Creationist author of *Genesis Flood*, singled out a "notorious Darwinian philosopher Michael Ruse," apparently a well-known "atheistic humanist," as contributing to the moral rot destroying contemporary America. "It is rather obvious that the modern opposition to capital punishment for murder and the general tendency toward leniency in punishment for other serious crimes are directly related to the strong emphasis on evolutionary determinism that has characterized much of this century."[4] I may disagree with this argumentation—I do, regretting only that I see no great general tendency toward leniency in punishment (especially capital punishment)—but at least I am not surprised. I expect this sort of thing from Creationists. Why then am I getting stuff of the same ilk from my fellow evolutionists? Goodness, I am even more hard-line than they always are. I think, for instance, that the Dawkins-Dennett enthusiasm for memes (units of culture) rather than genes (units of biology) is just plain silly. I want to go biological all the way down—or up.

At least part of the reason for the hostility is that, unlike these folk, I think that reasonable people can embrace science *and* religion, Darwinian evolutionary theory *and* Christianity. It is not something I want to do, but I do defend the integrity and possibility of others doing so. In part my case is political. If Darwinian theory leads to atheism (or agnosticism), then it seems to me that it is something with religious consequences. And I am not sure of the legality of teaching such stuff in state-supported schools in America. If the constitutional separation of Church and State keeps out Christianity then I suspect that it keeps out Darwinian theory also. I should say that this is only part of my case. If I thought the connection did hold, then I would feel obliged to follow it through to the end. But on intellectual grounds I think that a plausible argument can be made for embracing both Darwinian theory and Christianity, where this latter is taken in a traditional sense (which, among other things, repudiates naive biblical literalism).

Whether or not I am right in my arguments—read my *Can a Darwinian Be a Christian?* and judge for yourself—it has upset the militant atheistic Darwinians like Dawkins and Dennett and Coyne. They want no truck with the enemy—meaning religion of any kind—and they want no truck with anyone who is prepared to hold out a hand to the enemy, even though that person is a hard-line Darwinian and openly allows that he himself has no personal religious beliefs. I think of myself as a skeptic rather than an atheist—I prefer the term *skeptic* to *agnostic*, in part because it puts me in the tradition of David Hume and in part because so often agnostics are really not that interested in religion, and I am very—but if truth be known, I am pretty atheistic about most claims of Christianity. All-loving, all-powerful God; Jesus as the son of God; resurrection; eternal salvation—I don't believe in any of these. But none of this cuts any ice with those who have been categorized as the "new atheists." (For myself, I cannot see much new, but no matter.) Religion is evil and people who are not absolutist on these sorts of things are themselves evil. It is as simple as that. (I will refrain from adding: It is as simplistic as that.)

Thinking about this nasty little flare up has more deeply confirmed one belief that I have long held. Uncomfortable though it may be to admit, the Creationists are right in one of their claims about evolutionists. There is a strong tendency among evolutionists to turn their thinking into a form of secular religion. The dispute between me and the new atheists is very much like those sorts of disputes that plague Protestant churches, particularly those of an evangelical or conservative type. Things are going along nicely and then one group decides that the others are not sufficiently hard line on something—homosexuality, abortion, female priests, perhaps even the meaning of the Eucharist—and all hell lets loose. Instead of fighting the common enemy (Catholicism, Islam, atheism), all energies get directed toward anathematizing those with whom one shares 95 percent of one's beliefs. This is not proof that evolution is (or can be) a form of secular religion, but this is a consequence one would expect if it is.

This being so, let us devote this essay to the basic question: Why would one want to say that belief in evolution is (or can be) a form of secular religion? What are the grounds for saying something like this? As a deeply committed evolutionist myself, I always think that the answers to important questions lie in the past, so let us go back in time and work our way to the present.

BEFORE THE *ORIGIN*

Evolution is an idea with its roots in the eighteenth century. This was the time when the ideology of progress—the belief that humans through their own unaided efforts can change and improve their lots—became dominant, and there were several who took the cultural idea of progress and read it into the rocks, thereby making for an evolutionary or transmutationary view of life's history. Usually, they then promptly took their evolutionism and argued in a good circular fashion that this justified their beliefs in progress! Entirely typical was Charles Darwin's grandfather, Erasmus Darwin. Much taken by the fossils uncovered when once he took a trip into the cuttings for a new canal tunnel—"I have lately travel'd two days journey into the bowels of the earth, with three most able philosophers, and have seen the Goddess of Minerals naked, as she lay in her inmost bowers"[5]—Erasmus Darwin moved readily from a belief in social and industrial progress to one of progress in the organic world. Evolution from the primitive to the complex, from "monad to man" as the popular phrasing had it. "Would it be too bold to imagine, that all warm-blooded animals have arisen from one living filament, which THE GREAT FIRST CAUSE endued with animality, with the power of acquiring new parts, attended with new propensities, directed by irritations, sensations, volitions, and associations; and thus possessing the faculty of continuing to improve by its own inherent activity, and of delivering down those improvements by generation to its posterity, world without end?"[6]

Happy always to express his sentiments in florid verse, Erasmus Darwin broke forth on the joys and triumphs of evolutionary progress. Above all, there was the special position of humankind:

> Imperious man, who rules the bestial crowd,
> Of language, reason, and reflection proud,
> With brow erect who scorns this earthy sod,
> And styles himself the image of his God;
> Arose from rudiments of form and sense,
> An embryon point, or microscopic ens![7]

Not that we should think that humans succeed by force or superior senses of sight or touch or whatever. It is our reason that counts, together with other related organs like our hands:

> Proud Man alone in wailing weakness born,
> No horns protect him, and no plumes adorn;

No finer powers of nostril, ear or eye,
Teach the young Reasoner to pursue or fly.———
Nerved with fine touch above the bestial throngs,
The hand, first gift of Heaven! to man belongs.[8]

And this is all the end product of the progressive development of our intelligences, that cause and are reflected in our scientific achievements:

How loves and tastes, and sympathies commence
From evanescent notices of sense;
How from yielding touch and rolling eyes
The piles immense of human science rise![9]

Now I am not really interested here in the causal speculations that Erasmus Darwin advanced for his evolutionism. In fact, he rather favored some kind of inheritance of acquired characteristics—the process that came to be known as Lamarckism, after the French evolutionist who wrote a decade or two later than Erasmus Darwin. And I am certainly not in any sense implying that he was putting forward a theory that he thought of as atheistical, not merely going against Christianity but against a god of any kind. In fact, like many intellectuals at the end of the eighteenth century (including the first presidents of the United States), Erasmus Darwin was a deist—denying the Trinity and believing in God as an unmoved mover. The point rather was that he saw his evolutionism as part and parcel of his religious position, the one reinforcing and in turn being reinforced by the other. He saw the highest mark of God's power and glory not in divine interventions, miracles, but precisely in the fact that these are not needed. God can do everything through unbroken law. Evolution, the apotheosis of unbroken law, can therefore be seen as the ultimate climax of the divine creation—the proof above all else of God's standing and worth. In Darwin's own words: "What a magnificent idea of the infinite power of THE GREAT ARCHITECT! THE CAUSE OF CAUSES! PARENT OF PARENTS! ENS ENTIUM!"[10]

What is significant for our tale is that critics of evolution saw what was going on, and criticized it for this very reason. By the beginning of the nineteenth century sophisticated thinkers were moving on beyond a crude biblical literalism—the fossils and the rocks persuaded people that the history of the world is too long to be constrained by the genealogies of Genesis and that the denizens of the past are too varied and magnificent simply to be drowned in the Deluge—but they could see exactly why it would be unacceptable to a Christian. For the Protestant particularly, the hope of salvation lies in and only in God's unmerited grace. We are unworthy, but God takes pity on us and offers us the hope of eternal life. Providence is merciful. Evolution, with its backbone of progress, suggested that humans through their own unaided efforts can improve things. And this is heresy. Hence, although the critics of evolution certainly went after what they thought of as the scientific errors in evolutionism—gaps in the fossil record and that sort of thing—in respects they were more concerned to counter what they saw as its unacceptable, anti-Christian philosophy. The father of comparative anatomy, the great Frenchman George Cuvier (a Protestant despite his nationality) was typical. He savaged people like Lamarck on straight scientific grounds—how could one argue for evolution when the mummified animals brought back by Napoleon from Egypt are exactly the same species as those living and thriving today? But then he went after them as irreligious. How can one be an evolutionist and thus ignore—because one is thus appealing to a causal process governed by blind, undirected law—the evidence of God's hand in the creation? How can one ignore the functioning nature of organisms, as (with some justification) Cuvier felt that people like Lamarck were wont to do? For Cuvier, this was the key to understanding nature:

> Natural history nevertheless has a rational principle that is exclusive to it and which it employs with great advantage on many occasions; it is the *conditions of existence* or, popularly, *final causes*. As nothing may exist which does not include the conditions which made its existence possible, the different parts of each creature must be coordinated in such a way as to make possible the whole organism, not only in itself but in its relationship to those which surround it, and the analysis of these conditions often leads to general laws as well founded as those of calculation or experiment.[11]

And then, above all else, there is that dreadful appeal to progress, something that (as a Protestant) Cuvier abhorred, and as a servant of the state (and thinking that it was a philosophy that led to the Revolution) Cuvier feared. Against evolution, Cuvier saw no natural progress, no natural development of organisms. Rather, he supposed that every now and then organisms get wiped out by floods (of which, on historical grounds, Noah's may have been the last although not universal) and then they reinvade from other parts of the globe. "I do not pretend that a new creation was required for calling our present races of animals into existence. I only urge that they did not occupy the same places, and that they must have come from some other part of the globe."[12]

My point then is that early evolutionism was less a functioning empirical science—Erasmus Darwin was positively cavalier on empirical matters and Lamarck was not a whole lot better—and more a vehicle for a philosophy of deism-cum-progress. It was a challenge to Christianity, but for this reason rather than on strictly literalist grounds. Moreover, it was seen to be what it was. And this was a state of affairs that persisted until the middle of the nineteenth century. The best-known evolutionary tract was the *Vestiges of the Natural History of Creation*, authored anonymously by (whom we now know to be) the Scottish publisher Robert Chambers.[13] It was openly and flamboyantly progress endorsing, and backed by a Calvinist-deist view of the deity. And it was attacked in these terms. People like Adam Sedgwick, professor of geology at the University of Cambridge and a deeply committed, ardent Anglican, loathed and detested its message, thinking it the epitome of false religion—which of course from his perspective it was. At the same time, one should point out that there were those who loved the message of *Vestiges* precisely because of its message, one that either they held in its own right or that they managed to blend into their own readings of the gospel story. In an earlier essay (essay 7) we saw how the poet Alfred Tennyson incorporated many of Chambers's ideas into his famous and much-loved poem *In Memoriam*. At the end of the 1840s Tennyson read Chambers, or at least he read a very detailed review of Chambers's *Vestiges*. Chambers argued for an organic evolution that was unambiguously progressionist, that is to say, moving from simple forms up to humans and then perhaps beyond. Inspired by this, Tennyson picked up his pen and finished the poem. He argued in the final lines that perhaps there is meaning after all, despite a Lyellian uniformitarianism: that life is progressing upward, and that perhaps will go on beyond the human form that we have at present. Could it not be that Hallam represented some anticipation of the more-developed life to come, cut short, as it were, in its prime? There is therefore hope for us all and a meaning for the life of Hallam.

CHARLES DARWIN AND THE *ORIGIN OF SPECIES*

So much for evolution before 1859, the year in which the *Origin of Species* was published. What did Darwin do and how did he alter things? Start with what he did. (We can go quickly here because the main claims of the *Origin* were laid out in more detail in essay 1). Darwin set out to give a new theory of evolution, one that could indeed stand muster against a proper empirical approach to science. He made the fact of evolution secure and he proposed the mechanism, natural selection, that is today by scientists generally considered the key factor behind the development of organisms—a development by a slow natural process from a few simple forms, and perhaps indeed ultimately from inorganic substances (although, sagely, Darwin said nothing on this latter topic). In the *Origin*, after first stressing the analogy between the world of the breeder and the world of nature, and after showing how much variation exists between organisms in the wild, Darwin was then ready for the key inferences. First, an argument for the struggle for existence and, following on this, an argument for the mechanism of natural selection. "This preservation of favourable variations and the rejection of injurious variations, I call Natural Selection."[14] (For more details, see essay 5.)

With the main mechanisms of change thus presented, Darwin introduced the famous metaphor of a tree: "The affinities of all the beings of the same class have sometimes been represented by a great tree. I believe this simile largely speaks the truth." The leaves and twigs at the top represent the species extant today. Then as we go down the branches, we have the great evolutionary paths of yesterday. All the way down we go until we reach the very first shared origins of life. "As buds give rise by growth to fresh buds, and these, if vigorous, branch out and overtop on all sides many a feebler branch, so by generation I believe it has been with the great Tree of Life, which fills with its dead and broken branches the crust of the earth, and covers the surface with its ever branching and beautiful ramifications."[15]

Then from this, Darwin turned to a general survey of the biological world, offering what the philosopher William Whewell (1840) had dubbed a "consilience of inductions."[16] Each area was explained by evolution through natural selection and in turn each area contributed to the support of the mechanism of evolution through natural selection. Geographical distribution (biogeography) was paradigmatic, as Darwin explained just why it is that one finds the various patterns of animal and plant life around the globe. Another area where Darwin felt very confident was embryology. Why are the embryos of some different species very similar—humans and dogs, for instance—whereas the adults are very different? Darwin argued that this follows from the fact that in the womb the selective forces on the two embryos would be very similar—they would not therefore be torn apart—whereas the selective forces on the two adults would be very different—they would be torn apart. And finally, all of this led to that famous passage at the end of the *Origin*: "There is a grandeur in this view of life, with its several powers, having been originally breathed into a few forms or into one; and that, whilst this planet has gone cycling on according to the fixed law of gravity, from so simple a beginning endless forms, most beautiful and most wonderful have been, and are being, evolved."[17]

So much for the theory. (For more details, see essay 1.) Now, in the light of the history thus far presented, what was Darwin hoping to do? Two things we can say immediately: He was not repudiating progress. It may have had a somewhat subdued role, but as the quotation just given at the end of the last paragraph shows unambiguously, biological progress was there and believed in. And as later works like the *Descent of Man* showed very well, this biological progress was much bound up with a general belief in social and cultural progress:

> In all civilized countries man accumulates property and bequeaths it to his children. So that the children in the same country do not by any means start fair in the race for success. But this is far from an unmixed evil; for without the accumulation of capital the arts could not progress; and it is chiefly through their power that the civilised races have extended, and are now everywhere extending, their range, so as to take the place of the lower races.[18]

More than this. It is pretty clear that at the time of the writing of the *Origin*, Darwin subscribed to a deistic view of nature and the universe. To his American friend, Asa Gray, Harvard botanist and ardent Christian, Darwin wrote:

> With respect to the theological view of the question; this is always painful to me.—I am bewildered.—I had no intention to write atheistically. But I own that I cannot see, as plainly as others do, and as I should wish to do, evidence of design and beneficence on all sides of us. There seems to me too much misery in the world. I cannot persuade myself that a beneficent and omnipotent God would have designedly created the Ichneumonidae with the express intention of their feeding within the living bodies of caterpillars, or that a cat should play with mice. Not believing this, I see no necessity in the belief that the eye was expressly designed. On the other hand I cannot anyhow be contented to view this wonderful universe and especially the nature of man, and to conclude that everything is the result of brute force. I am inclined to look at everything as resulting from designed laws, with the details, whether good or bad, left to the working out of what we may call chance. Not that this notion *at all* satisfies me. I feel most deeply that the whole subject is too profound for the human intellect. A dog might as well speculate on the mind of Newton.—Let each man hope and believe what he can.[19]

And indeed, we can point to the Christian influences on Darwin's thinking. Adaptation, the phenomenon for which natural selection is supposed, is something that had been stressed by natural theologians from the time of Aristotle. Darwin got a full blast as a student when at Cambridge and he had to read Archdeacon Paley's *Natural Theology*, with the eye taken as a proof of the existence of God: "I know no better method of introducing so large a subject, than that of comparing a single thing with a single thing: an eye, for example, with a telescope."[20] Then as he moved to natural selection, Darwin grasped that artificial selection can bring about an equivalent of adaptation. The question was how artificial selection was to be transferred to the wild. It was when Darwin read Malthus, who argued that food and space supplies are exhausted by potential population growth and hence there is a struggle for existence, that Darwin then realized that he had in his grasp the missing part of his theory. But note that Malthus's position, although it sounds harsh (it is harsh!) is cast in a Christian framework. Without the spur of the struggle, no one would be inclined to work. God therefore put it in place deliberately for our own good.[21]

So I am certainly not saying that Darwin broke absolutely with his past. Indeed, in a way I am hinting that if someone were a Christian perhaps, for the first time, here was an evolutionary theory that might be molded and adapted for use without giving up one's faith. But, in the context of this present discussion, I think more important than the continuities was Darwin's determination to make of his theory something with a different status from those of his predecessors. Darwin did not want to produce a secular religion. He wanted to produce a functioning, empirical science. He wanted something, to use the language of Thomas Kuhn,[22] that could work as a "paradigm," making possible normal science. The kind of normal science that in fact he himself was to do soon after the *Origin*, when he wrote a little book on orchids[23] and that others were to do, like Henry W. Bates when he used natural selection to produce an explanation of butterfly mimicry.[24] Progress was there, but it was downplayed. References may have been made to the Creator, but He was given no work to do, and could have been dropped without loss of content. Evolution through natural selection was certainly going to contradict Genesis taken literally, but to think that Darwin was offering a "religion without revelation" (to borrow a title from a book of the twentieth century) would be quite to misunderstand his intent.

THE DARWINIAN EVANGELIST

So, what happened? I argue—and this is the most crucial point of this essay—that Darwinism got hijacked, and turned to other purposes. And the chief hijacker was none other than he who is celebrated as "Darwin's Bulldog," the nineteenth-century morphologist and paleontologist Thomas Henry Huxley. Unlike Darwin—a rich man, sick for most of his adult life, able to live as a semi-recluse—Huxley was a man who was making his own way, as a university professor and then as a college dean. He, with a number of others (mainly men but with some women like Florence Nightingale) were striving hard to change the course of British life, away from the near-feudalism of the rural eighteenth century and toward the modern, urban industrialism of the twentieth century. They were reforming the civil service, the military, the medical profession, and more—including teaching at school and university. Huxley was in the thick of creating a professional science—a professional science where one could succeed on merit and make a living—and Huxley realized full well that to achieve his aims he had to find reasons to employ the young scientists he was producing. Physiology he sold to the medical profession, arguing (with success) that the time had come to stop killing people and to start curing. Morphology he sold to the teaching profession—something at a crucial point, for only now was education starting to become the birthright of all and not under the sway of organized religion. For evolution, alas, Huxley could see no immediate cash value. It cured no pains in the belly, and it was too daring for the junior classroom. But Huxley—a dedicated evolutionist, albeit somewhat indifferent to natural selection—could nevertheless see a role for evolution. It would be the ideology—the secular religion—of the reformers, being something to put against the ideology—the spiritual religion—allied with those who resisted change. It would be the system giving answers to origins and explaining the status of humankind to replace the outdated system of the conservatives and reactionaries, who worshipped each Sunday in the local Anglican parish church. Evolution *versus* Christianity.

Progress, naturally, was to be the backbone of the system. But more was needed. A good religion has a moral system, a set of ethical prescriptions: Thou shalt not kill, Love your neighbor as yourself, and that sort of thing. Charles Darwin was not really into this sort of thing, but there was another English evolutionist ready and very willing to step into the breach. Herbert Spencer's evolutionism starts (continues and finishes) with progress. For him, progress was not so much an empirical finding but a metaphysical presupposition of his view of history. It ran through everything, from the most primitive forms of culture to the evolution of our own species:

> Now, we propose in the first place to show, that this law of organic progress is the law of all progress. Whether it be in the development of the Earth, in the development of Life upon its surface, in the development of Society, of Government, of Manufactures, of Commerce, of

Language, Literature, Science, Art, this same evolution of the simple into the complex, through successive differentiations, holds throughout. From the earliest traceable cosmical changes down to the latest results of civilization, we shall find that the transformation of the homogeneous into the heterogeneous, is that in which Progress essentially consists.[25]

What about causes? Never that interested in natural selection in the biological world, Spencer showed an eclectic synthesis of German morphology and British thermodynamics, seasoned with a good dash of British nonconformist thinking on society and the desirable underlying economic forces, arguing (perhaps more metaphysically than empirically) that nature starts in a condition of uniformity—what he called "homogeneity"—and tends naturally to a condition of complexity—what he called "heterogeneity." Why should this be so? Apparently it follows directly from the fact that causality tends to be open ended, inasmuch as one cause leads to multiple effects, rather than many causes leading to one effect. There is always a kind of explosion or expansion outward, as the simple and uniform tends to the complex and diverse. This happens at all levels of the hierarchy—organisms, states, whatever. Something internal or external jogs or disturbs the state of being, and the multiplying causal process kicks in. More than this however, for as the process of complexification is occurring, there is a tendency to move upward to a higher level of existence. Life—everything—is rather like the incoming tide, set on its end. There are surges forward, followed by moments or periods of consolidation, then further surges forward, with overall gain happening over and over again. Disturbance leads to the attempt to move back to a state of rest, but the new state is never that of the old state—it is more heterogeneous, and higher. Overall, therefore, evolution can be described (as it came to be known) as an exemplification of "dynamic equilibrium."

Morality fit nicely into all of this. It is our obligation to preserve and to promote progress. Here there is a place for the struggle and selection. Even in 1852, some years before the *Origin* was published, Spencer speculated on selective effects showing themselves in the different natures and behaviors of the Irish and the Scots. He concluded that struggle and selection in society translates into extreme laissez-faire socioeconomics:

the state should stay out of the way of people pursuing their own self-interests and should not at all attempt to regulate practices or redress imbalances or unfairnesses. Libertarian license therefore is not only the way that things are but the way that they should be. In fact, Spencer was far from convinced that mid-Victorian Britain was a laissez-faire society, but this is what he hoped fervently that it would become:

> We must call those spurious philanthropists, who, to prevent present misery, would entail greater misery upon future generations. All defenders of a Poor Law must, however, be classed among such. That rigorous necessity which, when allowed to act on them, becomes so sharp a spur to the lazy and so strong a bridle to the random, these pauper's friends would repeal, because of the wailing it here and there produces. Blind to the fact that under the natural order of things, society is constantly excreting its unhealthy, imbecile, slow, vacillating, faithless members, these unthinking, though well-meaning, men advocate an interference which not only stops the purifying process but even increases the vitiation—absolutely encourages the multiplication of the reckless and incompetent by offering them an unfailing provision, and *discourages* the multiplication of the competent and provident by heightening the prospective difficulty of maintaining a family.[26]

In fact, matters were rather more complex than this. Just as Christians can differ morally in the name of their savior—the battalion padre preaching fire and brimstone while the Quaker embraces pacifism—so followers of Spencer (who tended, somewhat inaccurately, to be called "Social Darwinians") could differ in their prescriptions. Spencer himself was far from denying the worth of any individual charity. It was rather state-supported institutions of charity that he opposed. The same is true very much of his followers. John D. Rockefeller the first, the founder of Standard Oil and one of the notorious businessmen at the beginning of this century, was openly in favor of denying state interference: he spent much of his time opposing the federal government as it strove to break up the monopoly he had established over the distribution and sale of fuel oil. He justified himself in Darwinian terms, saying that the fit do and should survive. Yet from the beginning he had tithed himself and always gave deeply to charity. Likewise Spencer enthusiast Andrew Carnegie, founder of US Steel, who claimed that no rich man should die rich. He gave much to the founding of public libraries. Interestingly here we see the direct input of a kind of Darwinism. Carnegie was less interested in stressing the downside of laissez faire, the failure of the unfit, than in stressing the upside, the success of the fit. Public libraries were places where poor-but-gifted children could go and thereby improve themselves and raise themselves up in society.[27]

These are details. Fascinating details, but details nevertheless. The point I make is that Charles Darwin was both a success and a failure. He was a success inasmuch (and it is a very big "inasmuch") as he turned people to evolution. Before him, it had been a pseudoscientific idea, on a par with astrology or phrenology. (Interestingly, Chambers had started to write a book on phrenology—the science of brain bumps—and changed halfway through to writing a book on evolution.) After Darwin, evolution was common sense. He was a failure inasmuch (and you judge how big an "inasmuch" you think this to be) as he did not turn evolution into a functioning, professional science, with natural selection at its heart. Evolution was a raging success, but more in a bastardized Spencerian version, functioning less as a science and more as a secular religion. That was what the reformers like Huxley wanted and that was what the reformers like Huxley got. When Jesus died on the cross, there was no religion of Christianity. That was for St. Paul to create, and people have been arguing ever since about the relationship between that life and teachings of Jesus and the religion that St. Paul left behind. When Darwin wrote the *Origin*, there was no science of Darwinism. That was for Thomas Henry Huxley to create, and I argue that the relationship between the teachings of Darwin and the religion of Huxley was about as iffy as that between Jesus and Paul.

THE SYNTHETIC THEORY

The 1930s saw the coming of Mendelian genetics generalized to populations, and with this it was possible to build a new Darwinism, one based on selection and a new and thriving theory of heredity. And with the intellectual advances came a determination by the supporters of this "synthetic theory" (a synthesis of Darwin and Mendel) to produce a functioning, mature, professional science of evolution. More than this—the determination came to fruition. Thanks to people like the Russian-born Theodosius Dobzhansky in America and E. B. Ford and his school of ecological genetics in England, Darwin's dream was realized. Perhaps slowly at first, but then with gathering speed, a selection-based, experiment- and observation-driven science of evolution came into being. One can mention Dobzhansky's *Genetics and the Origin of Species*,[28] Ford's *Ecological Genetics*,[29] as well as ornithologist Ernst Mayr's *Systematics and the Origin of Species*,[30] mammalian paleontologist George Gaylord Simpson's *Tempo and Mode in Evolution*,[31] and botanist G. Ledyard Stebbins's *Variation and Evolution in Plants*.[32] But here is the fascinating point. Every one of these people, and indeed the theoretical population geneticists (notably the English R. A. Fisher[33] and the American Sewall Wright[34]) on which the synthetic theorists based their empirical studies, was drawn first to evolution because it was a secular religion! Or at least (especially in the cases of Fisher and Dobzhansky) because it had the makings of a revitalization of the Christianity to which they subscribed already. They liked the idea of progress and they liked the idea of evolution yielding moral prescriptions. They did not want to give up on the extrascientific side of evolutionism.

So what were they to do? They wrote two sets of books. One set, dead straight and professional, with nary a hint of progress and so forth. Then another set, based on the first set, with the mathematics removed (not much work here, to be candid), with a couple of chapters on progress, morality, and the American way (or whatever), and a disarming preface telling you that this is for the "general reader." Simpson, probably the brightest of the lot, was a paradigm. *Tempo and Mode in Evolution*[35] is so professionally prim and proper it is almost boring. The same is true of the revision, *The Major Features of Evolution*,[36] that appeared some nine years later. But in the middle came *The Meaning of Evolution*,[37] and if the title does not give away the secret, then the contents do. There were masses of stuff on progress and on the implications of all of this. (See essay 6 for more details.) In fact, Simpson ran through a large number of proposed criteria of biological improvement or worth: expansion of life, dominance, specialization, potential for future development, independence from the environment, control of the environment, complexity, general energy level, pre- and postnatal care, sophisticated nervous system, individualization, and more. Humans certainly do not come out top on all of these. For instance, Simpson thought that we humans are not very specialized. However, overall, we tend to score well, the very best in many cases—like dominance and pre- and postnatal care and nervous system and individualization. And this general consilience seems to have been enough to convince Simpson that progress, with humans at the top, was more than just a whim or conceit. Progress may not be entirely objective, but it was more than just one man's yearning.

Move to ethics. Simpson was absolutely and completely committed to the view that ethics is natural, in the sense of being produced by evolution. Simpson argued that biology, through the medium of selection, could produce something like an ethical sense, pointing out that success in the struggle for existence does not necessarily mean all-out warfare, but can demand sympathetic alliance with one's fellows. Ethics is natural also in having no justification or sanction outside of evolution. What has evolved is what you get. Simpson, who came from a fundamentalist Presbyterian family, was by nature always an intensely religious man. But his faith in an existent deity was nonexistent (in middle life he worshipped with the Unitarians), and he certainly thought there could be no divine or similar support for moral belief.

What of moral belief? What should I do? Whereas some evolutionists (such as Julian Huxley, the grandson of Thomas Henry Huxley) were in favor of large-scale public works and other state-funded projects, Simpson looked much more to the individual level. There were two major directives. First, there was the need to improve and promote knowledge—knowledge in itself, as a good. "The most essential material factor in the new evolution seems to be just this: knowledge, together, necessarily, with its spread and inheritance."[38] Then secondly we have personal responsibility, which leads to integrity and dignity. "The responsibility is basically personal and becomes social only as it is extended in society among the individuals composing the social unit. It is correlated with another human evolutionary characteristic, that of high individualization."[39] And so on and so forth. I hardly have to say that the valuing of responsibility and dignity and so forth was equally a function of the times and society within which Simpson lived. We are talking now of the years when the cold war was settling right into its long winter, when Soviet science was suffering under influential charlatans like Lysenko, and when issues of dictatorship and totalitarianism were all too fresh in people's memories and present in much of the world of the day.

There is quite a bit more, but my point by now is surely clear. Even a hundred years after the *Origin*, even after natural selection had been promoted to the core of a solid, functioning, professional evolutionary biology, evolutionists—the very best evolutionists—were still using their theory as a Christianity substitute (or, in the case of some, as a Christianity enhancer).

THE TWENTY-FIRST CENTURY

Where are we today? Let me make one thing absolutely clear, although I am sure that it will be ignored by the Creationists in their selective quoting. I argue strongly and strenuously that there is today a mature evolutionary biology—Darwin based, empirical, predictive, explanatory. It has felt and benefited from the full blast of the molecular revolution in biology, and it looks forward into this new century with great accomplishments, with powerful tools, and with anticipation of solving major problems old and new. I mention simply as illustration the incredible advances over the past two decades in the understanding of development and of how this is now being integrated into the evolutionary picture (so-called evo-devo). (See essay 8.) This evolutionary biology is not, by any stretch of the imagination, a secular religion, and those who quote me as saying that it is (or pretend that I have not mentioned and stressed its existence and importance) do me and evolutionary biology a grave disfavor.

But, given our history, you would expect more to the story, and indeed there is. I argue also that—in the tradition of Thomas Henry Huxley, Herbert Spencer, and G. G. Simpson, and in an important way going right back to Erasmus Darwin and the birth of evolutionism— there is another side that continues unabated today. And this side does use evolution as a secular religion. Some who play this game are, like Simpson, great evolutionary biologists in their own right. One thinks here of the distinguished Harvard entomologist and sociobiologist Edward O. Wilson, who has made major advances in our understanding of social behavior. He nevertheless is explicit in wanting to make more of his science than mere science. We saw this in the last essay. Remember the use he makes of evolution in his Pulitzer Prize–winning *On Human Nature*. Speaking of scientific materialism he tells us: "It presents the human mind with an alternative mythology that until now has always, point for point in zones of conflict, defeated traditional religion. Its narrative form is the epic: the evolution of the universe from the big bang of fifteen billion years ago through the origin of the elements and celestial bodies to the beginnings of life on earth." This we learn is "mythology," but "the final decisive edge enjoyed by scientific naturalism will come from its capacity to explain traditional religion, its chief competition, as a wholly material phenomenon."[40]

Like Spencer (a thinker whom Wilson admires greatly), over the years Wilson has offered all sorts of moral prescriptions—precisely what one would expect of a secular religion (and not of an objective, disinterested scientific theory)—most particularly about the need to preserve biodiversity and to cherish the plants of the world, especially those vanishing from the Brazilian rainforests (where Wilson has spent much of his professional life). And it will not surprise the reader to find that progress is the force and reason behind everything: "The overall average across the history of life has moved from the simple and few to the more complex and numerous. . . . Progress, then, is a property of the evolution of life as a whole by almost any conceivable intuitive standard, including the acquisition of goals and intentions in the behavior of animals."[41] For Wilson, as for Spencer and Simpson, progress confers value and hence it is our obligation to promote (or at least not hinder) the evolutionary process.

There are others who play or have played this game. William Hamilton, who discovered kin selection, is generally considered the evolutionary genius of the second half of the twentieth century. In the words of Richard Dawkins, "Those of us who wish we had met Charles Darwin can console ourselves: we may have met the nearest equivalent that the late twentieth century had to offer."[42] Listen to Hamilton on the family:

> One of the ways in which I think backing plus curbing of the hypocrisies of individualism will come about will be through a greater measure of *family* responsibility that political parties will see it as a necessary measure to impose: as an example, if a family wants to keep a particular vegetable baby alive, the family must pay for it. Similarly, if a

church objects to the alternative—letting the baby die—extra taxes to pay for the baby's special care will be required from specifically that church's coffers, backing its beliefs. None will be demanded from another church that agrees the baby should die. In general along such lines, it will be a great step in the equitable running of modern society if a sincerity tax comes to be imposed on all propaganda—what you say you believe in you must show you believe in through hard cash and sacrifice; as an example again, there should be no option but that your child attends the idealistic comprehensive school you say you believe in.

But why not start to discuss and make decisions and try experiences now? Just as I believe Kosovars should be allowed to practice Islam in Kosovo, in the heart of Christian Europe, if they want to (including even practice the purdah and female circumcision if they want that) but not including any right of unlimited increase (or at least providing very strong disincentive, in contrast to the recent supposedly liberal and yet embittered spirit of toleration), so I believe that other self-defined yet quite different groups of idealists should be allowed to practice a religion that includes parent-decided selective infanticide, provided this is done under good safeguards against cruelty. Through free choice in idealism this will become real, effective group selection on the cultural plane, hopefully no longer contaminated from its warlike predecessor out of biology.[43]

I confess that something like this makes me shudder somewhat, but that is not the point now. It is rather that this appeared in a collection of Hamilton's papers on evolutionary theory. (Check the title: *The Narrow Roads of Gene Land*.) Also interesting and pertinent is the fact that the kinds of prescriptions that Hamilton is offering—not the content but the kinds—are not chosen randomly. Family values are the very things that religions care about most, morally. Think of the Old Testament and the constant harping on about the family. Think of the New Testament and of the scrambling of the early Christians to pretend that Jesus did not mean what he said when he told his followers to leave their families. Think of what obsesses so many evangelical Christians today—homosexuality, abortion, premarital sex, all things seen as threatening to family values. Phillip Johnson, the leader of the so-called Intelligent Design Movement and respected in his circles as highly as Hamilton in his, focuses right in on family values:

A responsible society is based first and foremost on responsible parents who fulfill their obligations to each other and to their children. Probably the most important thing that most adults do is to prepare the next generation for the joys and responsibilities of life. To do this they must ensure to the best of their ability that their children are born healthy. Following birth, children must be nurtured and educated in moral behavior by loving parents, preferably *two* parents. That is one reason it is important for lovers to regard marriage as a sacred bond, rather than as a contractual arrangement to be terminated at the convenience of either party. That is also why mothers in a rational society regard their children, born and unborn, as a sacred trust rather than primarily as an encumbrance that men impose on women in order to make them unhappy and impede their pursuit of wealth, power and pleasure. Similarly, fathers in a rational society regard their offspring from the beginning of pregnancy as their own flesh, so that they become enthusiastic providers and conurturers rather than the unwilling objects of child-support orders.[44]

I want to say that this is all part and parcel of Johnson's religion—it is the moral prescription part. I do not want to say that all moral prescriptions are necessarily religious. I make them and I would like to think that I am not religious. But it does seem to me that when you find the kinds of gaps that exist so evidently between Hamilton's evolutionary biology and his prescriptions, and when you find that he nevertheless thinks that the one follows from the other, it is not unfair to suspect that there is an underlying ideology, a world picture, that is influencing his thinking. And I do not see why I should not refer to this as a secular religion.

RICHARD DAWKINS

Now drawing to the end, let me return to the new atheists, Dawkins and company. Are they religious? They are always denying that this is so, and I suppose in the end it is a matter of semantics. If you like, let us say "religious-like." Consider Dawkins. He has characterized his move to atheism from religious belief as a "road to Damascus" experience,[45] and religious or not he certainly shows the fanaticism of the true believer. He and Saint Paul are peas in a pod. He believes that evolution leads to a philosophy of nihilism:

The universe we observe has precisely the properties we should expect if there is, at bottom, no design, no purpose, no evil and no good, nothing but blind, pitiless indifference. As that unhappy poet A. E. Houseman put it:

For Nature, heartless, witless Nature

Will neither know nor care.

DNA neither knows nor cares. DNA just is. And we dance to its music.[46]

Dawkins sees Darwinism as the intellectual alternative to Christianity:

Paley's argument is made with passionate sincerity and is informed by the best biological scholarship of his day, but it is wrong, gloriously and utterly wrong. The analogy between the telescope and the eye, between watch and living organism, is false. All appearances to the contrary, the only watchmaker in nature is the blind forces of physics, albeit deployed in a very special way. A true watchmaker has foresight: he designs his cogs and springs, and plans their interconnections, with a future purpose in his mind's eye. Natural selection, the blind, unconscious, automatic process which Darwin discovered, and which we now know is the explanation for the existence and apparently purposeful form of all life, has no purpose in mind. It has no mind and no mind's eye. It does not plan for the future. It has no vision, no foresight, no sight at all. If it can be said to play the role of watchmaker in nature, it is the *blind* watchmaker.[47]

Dawkins hates the beliefs of religious people. "The God of the Old Testament is arguably the most unpleasant character in all fiction: jealous and proud of it; a petty, unjust, unforgiving control-freak; a vindictive, bloodthirsty ethnic cleanser; a misogynistic, homophobic, racist, infanticidal, genocidal, filicidal, pestilential, megalomaniacal, sadomasochistic, capriciously malevolent bully."[48] He is not much friendlier to the God of the New Testament either, writing of "his insipidly opposite Christian face, 'Gentle Jesus meek and mild.'" Those are just the beliefs. Dawkins thinks the arguments of the other side are so bad that he cannot bring himself to read the literature and (as a consequence) presents schoolboy caricatures of the opposition. If you doubt this, look at the discussion of the arguments for the existence of God in *The God Delusion*. It is the first time in my life that I have felt sorry for the ontological argument. My favorite piece of atrocious reasoning is when Dawkins seems to think he has made a great discovery in showing that causal arguments end with God and then there is the question: What caused God? He has absolutely no knowledge of the vast literature that deals of God's existence and of the sense in which we can speak of Him as a necessary being. This notion, the notion of "aseity," may not work—I am not sure it does—but if you are going to write a whole book against a position and you proudly don't look at the other's arguments, one wonders about ideological agendas.

Finally, and most importantly, there is the fact that Dawkins is engaged in a moral crusade, not as a philosopher trying to establish premises and conclusions but as a preacher, telling the ways to salvation and to damnation. *The God Delusion* is above all a work of morality. Clearly fired by the horrors of 9/11, Dawkins is on a mission to crush religious belief in the name of science, in the name of Darwinian evolution. What does one say about someone who thinks that bringing children up in a Christian home is akin to child abuse? "Once, in the question time after a lecture in Dublin, I was asked what I thought about the widely publicized cases of sexual abuse by Catholic priests in Ireland. I replied that, horrible as sexual abuse no doubt was, the damage was arguably less than the long-term psychological damage inflicted by bringing the child up Catholic in the first place."[49]

As I said, perhaps in the end it is a matter of semantics. But no amount of word juggling is going to convince me that this is the voice of a scientist and nothing more.

ENVOI

Am I letting down the side? Today, in America particularly, we are in a battle between the forces of reason—the successors of the Enlightenment, the creators and lovers of science, the people of tolerance and understanding—and the forces of unreason, those who would take us back to the Middle Ages and earlier with their crude readings of scripture and their neofascist prescriptions for the ways in which we should all live our lives. Dawkins may be over the edge, but he is on our side. And after 9/11 should we not all be a little over the edge? Hamilton may have had ideas for social engineering that would have made the people of *Brave New World* take pause and think, but he was the really great evolutionist of his generation. Let us not bedaub him with dirt because of his silly social beliefs. Dennett may be extreme in likening religion to the liver fluke but he is the man who, in another context, referred to natural selection as the "greatest idea ever."[50] Don't pull down such a spokesman for the cause. Assuming that everything I have said in this essay is true, should I be bringing it into the light? Remember the wife of the Bishop of Worcester after the *Origin* was published: "Descended from monkeys? Let us hope that it is not true, but if it is true, let us hope it will not become widely known."

Basically I am going to leave this conundrum as an exercise for the reader. Personally, I think we do evolution no great favors by hiding the truth. Although I do not particularly want to do so myself, I do not mind if people want to use evolution as the basis for a kind of secular religion. I certainly prefer some forms (Wilson's) over others (Hamilton's). But each to his or her taste. For me, having given up the religion of my childhood, I am not very keen on embracing another religion for my old age. What I do say is that if some treat evolution as a religion, it is important to recognize this fact and to keep it separate from evolution as science. The Creationists want to attack all forms of evolution. Let us make sure that evolution as science—one of the noblest discoveries of the human mind—does not get dragged down and mired in ideological and constitutional mud because we are too frightened to admit that some evolutionists want to go beyond the science, trying to create alternatives for the lost beliefs of their childhoods.

NOTES

1. R. Dawkins, *The God Delusion* (New York: Houghton Mifflin, 2006).

2. D. Dennett, "Trapped in the Creationist Briar Patch," *Guardian*, April 4, 2006.

3. J. Coyne, "Intergalactic Jesus," *London Review of Books* 24 (May 9, 2002): 23–24.

4. H. M. Morris, *The Long War Against God: The History and Impact of the Creation/Evolution Conflict* (Grand Rapids, MI: Baker Book House, 1989), p. 148.

5. Letter to Josiah Wedgwood, July 2, 1767, in D. King-Hele, *The Letters of Erasmus Darwin* (Chicago: University of Chicago Press, 1981), p. 43.

6. E. Darwin, *Zoonomia; or, The Laws of Organic Life*, 3rd ed. (London: J. Johnson, 1801), 2:240.

7. E. Darwin, *The Temple of Nature* (London: J. Johnson, 1803), canto I, lines 309–14.

8. Ibid., canto III, lines 117–22.

9. Ibid., canto III, lines 43–46.

10. E. Darwin, *Zoonomia*, 2:247.

11. G. Cuvier, *Le règne animal distribué d'aprés son organisation, pour servir de base à l'histoire naturelle des animaux et d'introduction à l'anatomie comparée* (Paris: 1817), 1:6; quoted in W. Coleman, *Georges Cuvier Zoologist. A Study in the History of Evolution Theory* (Cambridge, MA: Harvard University Press, 1964), p. 42.

12. G. Cuvier, *Theory of the Earth*, 4th ed., ed. Robert Jameson (Edinburgh: William Blackwood, 1813), pp. 125–26.

13. R. Chambers, *Vestiges of the Natural History of Creation* (London: Churchill, 1844).

14. C. Darwin, *On the Origin of Species by Means of Natural Selection, or the Preservation of Favoured Races in the Struggle for Life* (London: John Murray, 1859), p. 81.

15. Ibid., pp. 129–30.

16. W. Whewell, *The Philosophy of the Inductive Sciences*, 2 vols. (London: Parker, 1840).

17. C. Darwin, *On the Origin of Species*, p. 490.

18. C. Darwin, *The Descent of Man, and Selection in Relation to Sex* (London: John Murray, 1871), 1:169.

19. Letter to Asa Gray, May 22, 1860, *The Correspondence of Charles Darwin* (Cambridge: Cambridge University Press, 1985–), 8: 224.

20. W. Paley, *Natural Theology*, Collected Works: IV (London: Rivington, 1819 [1802]), p. 14.

21. T. R. Malthus, *An Essay on the Principle of Population*, 6th ed. (London: Everyman, 1914 [1826]).

22. T. Kuhn, *The Structure of Scientific Revolutions* (Chicago: University of Chicago Press, 1962).

23. C. Darwin, *On the Various Contrivances by which British and Foreign Orchids Are Fertilized by Insects, and On the Good Effects of Intercrossing* (London: John Murray, 1862).

24. H. W. Bates, "Contributions to an Insect Fauna of the Amazon Valley," *Transactions of the Linnean Society of London*, 1862.

25. H. Spencer, "Progress: Its Law and Cause," *Westminster Review* 67 (1857): 446–47.

26. H. Spencer, *Social Statics; or, The Conditions Essential to Human Happiness Specified and the First of Them Developed* (London: J. Chapman: 1851), pp. 323–24.

27. R. Bannister, *Social Darwinism: Science and Myth in Anglo-American Social Thought* (Philadelphia: Temple University Press, 1979); C. E. Russett, *Darwin in America: The Intellectual Response. 1865–1912* (San Francisco: Freeman, 1976).

28. T. Dobzhansky, *Genetics and the Origin of Species* (New York: Columbia University Press, 1937).

29. E. B. Ford, *Ecological Genetics* (London: Methuen, 1964).

30. E. Mayr, *Systematics and the Origin of Species* (New York: Columbia University Press, 1942).

31. G. G. Simpson, *Tempo and Mode in Evolution* (New York: Columbia University Press, 1944).

32. G. L. Stebbins, *Variation and Evolution in Plants* (New York: Columbia University Press, 1950).

33. R. A. Fisher, *The Genetical Theory of Natural Selection* (Oxford: Oxford University Press, 1930).

34. S. Wright, "Evolution in Mendelian Populations," *Genetics* 16 (1931): 97–159; "The Roles of Mutation, Inbreeding, Crossbreeding and Selection in Evolution," *Proceedings of the Sixth International Congress of Genetics* 1 (1932): 356–66.

35. Simpson, *Tempo and Mode in Evolution*.

36. G. G. Simpson, *The Major Features of Evolution* (New York: Columbia University Press, 1953).

37. G. G. Simpson, *The Meaning of Evolution* (New Haven, CT: Yale University Press, 1949).

38. Ibid., p. 311.

39. Ibid., p. 315.

40. E. O. Wilson, *On Human Nature* (Cambridge, MA: Harvard University Press, 1978), p. 192.

41. E. O. Wilson, *The Diversity of Life* (Cambridge, MA: Harvard University Press, 1992), p. 187.

42. W. D. Hamilton, *Narrow Roads of Gene Land: The Collected Papers of W. D. Hamilton*, vol. 2, *Evolution of Sex* (Oxford: Oxford University Press, 2001), p. xi.

43. Ibid., p. xlviii.

44. P. E. Johnson, *Reason in the Balance: The Case Against Naturalism in Science, Law and Education* (Downers Grove, IL: InterVarsity Press, 1995), pp. 150–51.

45. R. Dawkins 1997.

46. R. Dawkins, "Richard Dawkins: A Survival Machine," in *The Third Culture*, ed. J. Brockman (New York: Simon and Schuster, 1995), pp. 131–33.

47. R. Dawkins, *The Blind Watchmaker* (New York: Norton, 1986), p. 5.

48. R. Dawkins, *God Delusion*, p. 31.

49. Ibid., p. 317.

50. D. C. Dennett, *Darwin's Dangerous Idea* (New York: Simon and Schuster, 1995).

PERMISSIONS

I am grateful for permission to publish here the following essays that first appeared elsewhere.

Ruse, M. 2004. "Adaptive Landscapes and Dynamic Equilibrium: The Spencerian Contribution to Twentieth-Century American Evolutionary Biology." In *Darwinian Heresies*, edited by A. Lustig, R. J. Richards, and M. Ruse, pp. 131–50. Cambridge: Cambridge University Press.

———. 2005. "Darwin and Mechanism: Metaphor in Science." *Studies in History and Philosophy of Biology and Biomedical Sciences* 36: 285–302.

———. 2005. "Evo-devo: A New Evolutionary Paradigm?" In *Philosophy, Biology, and Life (Royal Institute of Philosophy Supplements)*, edited by A. O'Heare. Cambridge: Cambridge University Press.

———. 2006. "Kant and Evolution." In *Theories of Generation*, edited by J. Smith. Cambridge: Cambridge University Press.

———. 2007. "Charles Darwin's *On the Origin of Species*." *Topoi* 26.

REFERENCE LIST

Agassiz, E. C., ed. 1885. *Louis Agassiz: His Life and Correspondence.* 2 vols. Boston: Houghton Mifflin.

Appel, T. A. 1987. *The Cuvier-Geoffery Debate: French Biology in the Decades Before Darwin.* New York: Oxford University Press.

Atran, S. 2004. *In Gods We Trust: The Evolutionary Landscape of Religion.* New York: Oxford University Press.

Bailey, L. H. 1897. *The Survival of the Unlike: A Collection of Evolution Essays Suggested by the Study of Domestic Plants.* 2nd ed. New York: Macmillan.

Balfour, F. M. 1880–1881. *A Treatise on Comparative Embryology.* London: Macmillan.

Bannister, R. 1979. *Social Darwinism: Science and Myth in Anglo-American Social Thought.* Philadelphia: Temple University Press.

Barnes, J., ed. 1984. *The Complete Works of Aristotle.* Princeton, NJ: Princeton University Press.

Bates, H. W. 1862. "Contributions to an Insect Fauna of the Amazon Valley." *Transactions of the Linnean Society of London* 23: 495–515.

Bergson, H. 1911. *Creative Evolution.* New York: Holt.

Bowler, P. 1984. *Evolution: The History of an Idea.* Berkeley: University of California Press.

Boyer, P. 2002. *Religion Explained: The Evolutionary Origins of Religious Thought.* New York: Basic Books.

Boyle, R. [1688] 1966. "A Disquisition about the Final Causes of Natural Things." In *The Works of Robert Boyle*, ed. T. Birch, 5:392–444. Hildesheim: Georg Olms.

———. 1996. *A Free Enquiry into the Vulgarly Received Notion of Nature*, ed. E. B. Davis and M. Hunter. Cambridge: Cambridge University Press.

Brewster, D. 1838. "Review of Comte's 'Cours de Philosophie Positive.'" *Edinburgh Review* 67: 271–308.

Broad, C. D. 1944. "Critical Notice of Julian Huxley's *Evolutionary Ethics.*" *Mind* 53.

———. 1971. "Critical Notice of Julian Huxley's *Evolutionary Ethics.*" (Reprinted from *Mind*, 1944). In *Broad's Critical Essays in Moral Philosophy*, ed. D. R. Cheney, 156–87. London: Allen and Unwin.

Browne, J. 1995. *Charles Darwin: Voyaging. Volume I of a Biography.* New York: Knopf.

———. 2002. *Charles Darwin: The Power of Place. Volume II of a Biography*. New York: Knopf.

Cannon, W. F. 1961. "The Impact of Uniformitarianism: Two Letters from John Herschel to Charles Lyell, 1836–1837." *Proceedings of the American Philosophical Society* 105: 301–14.

Carroll, S. B. 1995. "Homeotic Genes and the Evolution of Arthropods." *Nature* 376: 479–85.

Carroll, S. B., J. K. Grenier, and S. D. Weatherbee. 2001. *From DNA to Diversity: Molecular Genetics and the Evolution of Animal Design*. Oxford: Blackwell.

Chambers, R. 1844. *Vestiges of the Natural History of Creation*. London: Churchill.

Coleman, W. 1964. *Georges Cuvier Zoologist. A Study in the History of Evolution Theory*. Cambridge, MA: Harvard University Press.

Conklin, E. G. 1921. *The Direction of Human Evolution*. London: Oxford University Press.

Cornell, J. F. 1986. "Newton of the Grassblade? Darwin and the Problem of Organic Teleology." *Isis* 77: 405–21.

Coyne, J. A. 2002. "Intergalactic Jesus." *London Review of Books* 24 (May 9): 23–24.

Coyne, J. A., N. H. Barton, and M. Turelli. 1997. "Perspective: A Critique of Sewall Wright's Shifting Balance Theory of Evolution." *Evolution* 51, no. 3: 643–71.

Cuvier, G. [1813] 1822. *Theory of the Earth*, 4th ed., ed. Robert Jameson. Edinburgh: William Blackwood.

———. 1817. *Le règne animal distribué d'aprés son organisation, pour servir de base à l'histoire naturelle des animaux et d'introduction à l'anatomie comparée*. Paris.

Darwin, C. 1839. *Journal of Researches into the Geology and Natural History of the Various Countries Visited by HMS Beagle*. London: Henry Colburn. (Note: Modern editions often titled *The Voyage of the Beagle*.)

———. 1845. *Journal of Researches into the Natural History and Geology of the Countries Visited during the Voyage of H.M.S. Beagle round the World*, 2nd ed. London: John Murray.

———. 1854. *A Monograph of the Fossil Balanidae and Verrucidae of Great Britain*. London: Palaeontographical Society.

———. 1859. *On the Origin of Species by Means of Natural Selection, or the Preservation of Favoured Races in the Struggle for Life*. London: John Murray.

———. 1862. *On the Various Contrivances by which British and Foreign Orchids Are Fertilized by Insects, and On the Good Effects of Intercrossing*. London: John Murray.

———. 1868. *The Variation of Animals and Plants Under Domestication*. London: John Murray.

———. 1871. *The Descent of Man, and Selection in Relation to Sex*. London: John Murray.

———. 1872. *The Expression of the Emotions in Man and Animals*. London: John Murray.

———. 1969. *Autobiography*. New York: Norton.

———. 1985. *The Correspondence of Charles Darwin*. Cambridge: Cambridge University Press.

Darwin, C., and A. R. Wallace. 1958. *Evolution by Natural Selection*. Foreword by Gavin de Beer. Cambridge: Cambridge University Press.

Darwin, E. 1794–1796. *Zoonomia; or, The Laws of Organic Life*. London: J. Johnson.

———. 1801. *Zoonomia; or, The Laws of Organic Life*. 3rd ed. London: J. Johnson.

———. 1803. *The Temple of Nature*. London: J. Johnson.

Dawkins, R. 1976. *The Selfish Gene*. Oxford: Oxford University Press.

———. 1983. "Universal Darwinism." In *Molecules to Men*, ed. D. S. Bendall. Cambridge: University of Cambridge Press.

———. 1986. *The Blind Watchmaker*. New York: Norton.

———. 1995. "Richard Dawkins: A Survival Machine." In *The Third Culture*, ed. J. Brockman. New York: Simon and Schuster.

———. 2006. *The God Delusion*. New York: Houghton Mifflin.

Dennett, D. C. 1995. *Darwin's Dangerous Idea*. New York: Simon and Schuster.

———. 2006. *Breaking the Spell: Religion as a Natural Phenomenon*. New York: Viking.

———. April 4, 2006. "Trapped in the Creationist Briar Patch." *Guardian*.

Dobzhansky, T. 1937. *Genetics and the Origin of Species*. New York: Columbia University Press.

Dreiser, T. 1900. *Sister Carrie*. New York: Doubleday, Page.

Duncan, D., ed. 1908. *Life and Letters of Herbert Spencer*. London: Williams and Norgate.

Eldredge, N., and S. J. Gould. 1972. "Punctuated Equilibria: An Alternative to Phyletic Gradualism." In *Models in Paleobiology*, ed. T. J. M. Schopf, pp. 82–115. San Francisco: Freeman, Cooper.

Fichman, M. 2004. *An Elusive Victorian: The Evolution of Alfred Russel Wallace*. Chicago: University of Chicago Press.

Fisher, R. A. 1930. *The Genetical Theory of Natural Selection*. Oxford: Oxford University Press.

Fodor, J. 1996. "Peacocking." *London Review of Books*, no. 18 (April): 19–20.

Ford, E. B. 1964. *Ecological Genetics*. London: Methuen.

Foucault, M. 1970. *The Order of Things: An Archaeology of the Human Sciences*. New York: Pantheon.

Freeman, S., and J. C. Herron. 2004. *Evolutionary Analysis*. 3rd ed. Englewood Cliffs, NJ: Prentice-Hall.

Freud, S. [1905] 1955. "Three Essays on the Theory of Sexuality." In *The Standard Edition of the Complete Psychological Works of Sigmund Freud*, 7:125–243. London: Hogarth.

Gayon, Jean. 1992. *Darwin et l'après-Darwin: Une histoire de l'hypothèse de sélection naturelle*. Paris: Kimé.

Gegenbaur, C. 1878. *Elements of Comparative Anatomy*. London: Macmillan.

Gilbert, S. F., J. M. Opitz, and R. A. Raff. 1996. "Resynthesizing Evolutionary and Developmental Biology." *Developmental Biology* 173: 357–72.

Gillespie, C. C. 1950. *Genesis and Geology*. Cambridge, MA: Harvard University Press.

Goodrich, E. S. 1912. *The Evolution of Living Organisms*. London: T. C. and E. C. Jack.

Gould, S. J. 2002. *The Structure of Evolutionary Theory*. Cambridge, MA: Harvard University Press.

Gould, S. J., and R. C. Lewontin. 1979. "The Spandrels of San Marco and the Panglossian Paradigm: A Critique of the Adaptationist Programme." *Proceedings of the Royal Society of London, Series B: Biological Sciences* 205: 581–98.

Grant, P. R. 1986. *Ecology and Evolution of Darwin's Finches*. Princeton, NJ: Princeton University Press.

Grant, R. B., and P. R. Grant. 1989. *Evolutionary Dynamics of a Natural Population: The Large Cactus Finch of the Galapagos*. Chicago: University of Chicago Press.

Gray, A. 1876. *Darwiniana*. New York: D. Appleton. (Note: Reprinted edition A. H. Dupree, ed. 1963. Cambridge, MA: Harvard University Press.)

Greene, J. C. *Science, Ideology and World View*. Berkeley: University of California Press, 1981.

Haeckel, E. 1866. *Generelle Morphologie der Organismen*. Berlin: Georg Reimer.

Haldane, J. B. S. 1931. "A Mathematical Theory of Natural and Artificial Selection. Pt. VIII. Metastable Populations." *Transactions of the Cambridge Philosophical Society* 27: 137.

———. 1932. *The Causes of Evolution*. New York: Cornell University Press.

Hamilton, W. D. 2001. *Narrow Roads of Gene Land: The Collected Papers of W. D. Hamilton*. Vol. 2, *Evolution of Sex*. Oxford: Oxford University Press.

Harré, R. 1972. *The Philosophies of Science: An Introductory Survey*. Oxford: Oxford University Press.

Hegel, G. W. F. [1817] 1970. *Philosophy of Nature*. Oxford: Oxford University Press.

Henderson, L. J. 1913. *The Fitness of the Environment*. New York: Macmillan.

———. 1917. *The Order of Nature*. Cambridge, MA: Harvard University Press.

Herschel, J. F. W. 1830. *Preliminary Discourse on the Study of Natural Philosophy.* London: Longman, Rees, Orme, Brown, Green, and Longman.

Hume, D. [1757] 1963. "A Natural History of Religion." In *Hume on Religion,* ed. R Wollheim. London: Fontana.

———. 1978. *A Treatise of Human Nature.* Oxford: Oxford University Press.

Huxley, J. S. 1912. *The Individual in the Animal Kingdom.* Cambridge: Cambridge University Press.

———. 1914. *The Courtship Habits of the Great Crested Grebe.* London: Jonathan Cape.

———. 1923. *Essays of a Biologist.* London: Chatto and Windus.

———. 1924. "The Negro Problem." *Spectator,* November 29: 821–22.

———. 1927. *Religion Without Revelation.* London: Ernest Benn.

———. 1931. *What Dare I Think? The Challenge of Modern Science to Human Action and Belief.* London: Chatto.

———. 1932. *The Captive Shrew and Other Poems of a Biologist.* Oxford: Basil Blackwell.

———. 1932. *Problems of Relative Growth.* London: Methuen.

———. 1934. *If I Were Dictator.* New York and London: Harper and Brothers.

———. 1942. *Evolution: The Modern Synthesis.* London: Allen and Unwin.

———. 1943. "TVA, an Achievement of Democratic Planning." *Architectural Review* 93, no. 558 (June): 138–66.

———. 1948. *UNESCO: Its Purpose and Its Philosophy.* Washington, DC: Public Affairs Press.

———. 1957. *New Bottles for New Wine.* London: Chatto and Windus.

———. 1964. *Essays of a Humanist.* London: Chatto and Windus.

———. 1970. *Memories.* London: Allen and Unwin.

———. 1973. *Memories II.* London: Allen and Unwin.

Huxley, L. 1900. *The Life and Letters of Thomas Henry Huxley.* 2 vols. London: Macmillan.

Huxley, T. H. 1873. *Critiques and Addresses.* New York: Appleton.

———. 1879. *Hume.* London: Macmillan.

———. 1893. "Evolution and Ethics." In *Evolution and Ethics,* 46–116. London: Macmillan.

Huxley, T. H., and J. S. Huxley. 1947. *Evolution and Ethics 1893–1943.* London: Pilot.

Huxley, T. H., and H. N. Martin. 1875. *A Course of Practical Instruction in Elementary Biology.* London: Macmillan.

James, W. 1902. *Varieties of Religious Experience: A Study in Human Nature.* New York: Longman.

Johnson, P. E. 1995. *Reason in the Balance: The Case Against Naturalism in Science, Law and Education.* Downers Grove, IL: InterVarsity Press.

Jones, G. 1980. *Social Darwinism and English Thought*. Brighton, UK: Harvester.

Kant, I. [1790] 1928. *The Critique of Teleological Judgement*, trans. J. C. Meredith. Oxford: Oxford University Press.

———. 1993. *Opus Postumum*, ed. E Förster. Cambridge: Cambridge University Press.

Kellogg, V. L. 1907. *Darwinism Today*. New York: Henry Holt.

Kettlewell, H. B. D. 1973. *The Evolution of Melanism*. Oxford: Clarendon.

King-Hele, D., ed. 1981. *The Letters of Erasmus Darwin*. Cambridge: Cambridge University Press.

Kipling, R. 1914. *Kim*. Garden City, NY: Doubleday, Page.

Kuhn, T. 1962. *The Structure of Scientific Revolutions*. Chicago: University of Chicago Press.

———. 1993. "Metaphor in Science." In *Metaphor and Thought*. 2nd ed., ed. Andrew Ortony, pp. 533–42. Cambridge: Cambridge University Press.

Laporte, L. 2000. *George Gaylord Simpson*. New York: Columbia University Press.

London, J. 1903. *The Call of the Wild*. New York: Macmillan.

Lovejoy, A. O. [1911] 1959. "Kant and Evolution." In *Forerunners of Darwin*, ed. B. Glass, O. Temkin, and W. L. Strauss Jr., pp. 173–206. Baltimore: Johns Hopkins University Press.

Lurie, E. 1960. *Louis Agassiz: A Life in Science*. Chicago: Chicago University Press.

Lyell, C. 1830–1833. *Principles of Geology: Being an Attempt to Explain the Former Changes in the Earth's Surface by Reference to Causes Now in Operation*. London: John Murray.

MacCulloch, D. 2004. *The Reformation: A History*. New York: Viking.

Malthus, T. R. [1826] 1914. *An Essay on the Principle of Population*, 6th ed. London: Everyman.

Marchant, J., ed. 1916. *Alfred Russel Wallace: Letters and Reminiscences*. London: Cassell and Company, Ltd.

Maynard Smith, J. 1969. "The Status of Neo-Darwinism." In *Towards a Theoretical Biology*, ed. C. H. Waddington. Edinburgh: Edinburgh University Press.

Maynard Smith, J., R. Burian, S. Kauffman, P. Alberch, J. Campbell, B. Goodwin, R. Lande, D. Raup, and L. Wolpert. 1985. "Developmental Constraints and Evolution." *Quarterly Review of Biology* 60: 265–87.

Mayr, E. 1942. *Systematics and the Origin of Species*. New York: Columbia University Press.

———. 1982. *The Growth of Biological Thought: Diversity, Evolution and Inheritance*. Cambridge, MA: Harvard University Press.

———. 1988. *Towards a New Philosophy of Biology: Observations of an Evolutionist.* Cambridge, MA: Belknap.

Mayr, E., and W. Provine, eds. 1980. *The Evolutionary Synthesis: Perspectives on the Unification of Biology.* Cambridge, MA: Harvard University Press.

McEwan, I. 1997. *Enduring Love.* London: Cape.

McMullin, E. 1983. "Values in Science." *PSA 1982*, ed. P. D. Asquith and T. Nickles, pp. 3–28. East Lansing, MI: Philosophy of Science Association.

Moore, B. 1913. *The Origin and Nature of Life.* London: Williams and Norgate.

Moore, G. E. 1903. *Principia Ethica.* Cambridge: Cambridge University Press.

Morrell, J., and A. Thackray. 1981. *Gentlemen of Science: Early Years of the British Association for the Advancement of Science.* Oxford: Oxford University Press.

Morris, H. M. 1989. *The Long War Against God: The History and Impact of the Creation/Evolution Conflict.* Grand Rapids, MI: Baker Book House.

Murphy, J. 1982. *Evolution, Morality, and the Meaning of Life.* Totowa, NJ: Rowman and Littlefield.

Norris, F. 1901. *Octopus: A Story of California.* New York: Doubleday, Page.

Osborn, H. F. 1931. *Cope: Master Naturalist: The Life and Writings of Edward Drinker Cope.* Princeton, NJ: Princeton University Press.

Owen, R. 1848. *On the Archetype and Homologies of the Vertebrate Skeleton.* London: Voorst.

———. 1860. "Darwin on the *Origin of Species.*" *Edinburgh Review* 111: 487–532.

Owen, Rev. R. 1894. *The Life of Richard Owen.* London: Murray.

Paley, W. [1802] 1819. *Natural Theology* (Collected Works: IV). London: Rivington.

Peirce, C. S. 1982. *Pragmatism.* Indianapolis: Hackett.

Pittenger, M. 1993. *American Socialists and Evolutionary Thought, 1870–1920.* Madison: University of Wisconsin Press.

Popper, K. R. 1974. "Darwinism as a Metaphysical Research Programme." In *The Philosophy of Karl Popper.* Vol. 1, ed. P. A. Schilpp, pp. 133–43. LaSalle, IL: Open Court.

Poulton, E. B. 1890. *The Colours of Animals.* London: Kegan Paul, Trench, Truebner.

Powell, B. 1855. *Essays on the Spirit of the Inductive Philosophy.* London: Longman, Brown, Green, and Longmans.

Provine, W. B. 1986. *Sewall Wright and Evolutionary Biology.* Chicago: University of Chicago Press.

Pusey, J. R. 1983. *China and Charles Darwin.* Cambridge, MA: Harvard University Press.

Raby, P. 2001. *Alfred Russel Wallace: A Life.* Princeton, NJ: Princeton University Press.

Raff, R. 1996. *The Shape of Life: Genes, Development, and the Evolution of Animal Form*. Chicago: University of Chicago Press.

Ray, J. 1709. *The Wisdom of God, Manifested in the Words of Creation*, 5th ed. London: Samuel Smith.

Reeve, H. K., and P. W. Sherman. 1993. "Adaptation and the Goals of Evolutionary Research." *Quarterly Review of Biology* 68: 1–32.

Reynolds, V., and R. Tanner. 1983. *The Biology of Religion*. London: Longman.

Reznick, D. N., M. V. Butler IV, and H. Rodd. 1996. "Differential Mortality as a Mechanism for Natural Selection in the Guppy (*Poecilia reticulata*)." *Evolution* 50: 1651–60.

Reznick, D. N., and J. Travis. 1996. "The Empirical Study of Adaptation in Natural Populations." In *Adaptation*, ed. M. R. Rose and G. V. Lauder. San Diego: Academic Press.

Richards, R. J. 1987. *Darwin and the Emergence of Evolutionary Theories of Mind and Behavior*. Chicago: University of Chicago Press.

———. 1992. *The Meaning of Evolution: The Morphological Construction and Ideological Reconstruction of Darwin's Theory*. Chicago: University of Chicago Press.

———. 2003. *The Romantic Conception of Life: Science and Philosophy in the Age of Goethe*. Chicago: University of Chicago Press.

———. 2004. "Michael Ruse's Design for Living." *Journal of the History of Biology* 37: 25–38.

Roth, G., J. Blanke, and D. B. Wake. 1994. "Cell Size Predicts Morphological Complexity in the Brains of Frogs and Salamanders." *Proceedings of the National Academy of the Sciences, USA* 91: 4796–800.

Ruse, M. 1975. "Darwin's Debt to Philosophy: An Examination of the Influence of the Philosophical Ideas of John F. W. Herschel and William Whewell on the Development of Charles Darwin's Theory of Evolution." *Studies in History and Philosophy of Science* 6: 159–81.

———. 1979. *The Darwinian Revolution: Science Red in Tooth and Claw*. Chicago: University of Chicago Press.

———. 1980. "Charles Darwin and Group Selection." *Annals of Science* 37: 615–30.

———. 1982. *Darwinism Defended: A Guide to the Evolution Controversies*. Reading, MA: Benjamin/Cummings Pub. Co.

———. 1986. *Taking Darwin Seriously: A Naturalistic Approach to Philosophy*. Oxford: Blackwell.

———. 1989. "Is the Theory of Punctuated Equilibria a New Paradigm?" *Journal of Social and Biological Structures* 12: 195–212.

———. 1994. *Evolutionary Naturalism: Selected Essays*. London: Routledge.

———. 1996. *Monad to Man: The Concept of Progress in Evolutionary Biology.* Cambridge, MA: Harvard University Press.

———. 1999. *Mystery of Mysteries: Is Evolution a Social Construction?* Cambridge, MA: Harvard University Press.

———. 2000. "Robert Boyle and the Machine Metaphor." *Zygon* 37: 581–95.

———. 2001. *Can a Darwinian Be a Christian? The Relationship between Science and Religion.* Cambridge: Cambridge University Press.

———. 2003. *Darwin and Design: Does Evolution Have a Purpose?* Cambridge, MA: Harvard University Press.

———. 2004. "The Romantic Conception of Robert J. Richards." *Journal of the History of Biology* 37: 3–23.

———. 2005. *The Evolution-Creation Struggle.* Cambridge, MA: Harvard University Press.

———. 2006. *Darwinism and Its Discontents.* Cambridge: Cambridge University Press.

Ruse, M., and R. J. Richards, eds. 2008. *The Cambridge Companion to the "Origin of Species."* Cambridge: Cambridge University Press.

Ruse, M., and E. O. Wilson. 1985. "The Evolution of Morality." *New Scientist* 1478: 108–28.

Russell, E. S. 1916. *Form and Function: A Contribution to the History of Animal Morphology.* London: John Murray (Reprinted by the University of Chicago Press, 1982).

Russett, C. E. 1966. *The Concept of Equilibrium in American Social Thought.* New Haven, CT: Yale University Press.

———. 1976. *Darwin in America: The Intellectual Response. 1865–1912.* San Francisco: Freeman.

Secord, J. A. 2000. *Victorian Sensation: The Extraordinary Publication, Reception, and Secret Authorship of Vestiges of the Natural History of Creation.* Chicago: University of Chicago Press.

Shaw, G. B. 1988. *Back to Methuselah: A Metabiological Pentateuch.* Harmondsworth, UK: Penguin.

Shermer, M. 2002. *In Darwin's Shadow: The Life and Science of Alfred Russel Wallace.* New York: Oxford University Press.

Sidgwick, H. 1874. *The Methods of Ethics.* London: Macmillan.

Simpson, G. G. 1944. *Tempo and Mode in Evolution.* New York: Columbia University Press.

———. 1949. *The Meaning of Evolution.* New Haven, CT: Yale University Press.

———. 1953. *The Major Features of Evolution.* New York: Columbia University Press.

———. 1964. *This View of Life.* New York: Harcourt, Brace, and World.

———. 1978. *Concession to the Improbable: An Unconventional Autobiography*. New Haven, CT: Yale University Press.

Spencer, H. 1851. *Social Statics; or, The Conditions Essential to Human Happiness Specified and the First of Them Developed*. London: J. Chapman.

———. 1852. "A Theory of Population, Deduced from the General Law of Animal Fertility." *Westminster Review* 1: 468–501.

———. 1857. "Progress: Its Law and Cause." *Westminster Review* 67: 445–85.

———. 1862. *First Principles*. London: Williams and Norgate.

———. 1864. *Principles of Biology*. 2 vols. London: Williams and Norgate.

Stebbins, G. L. 1950. *Variation and Evolution in Plants*. New York: Columbia University Press.

Stebbins, G. L., and F. J. Ayala. 1981. "Is a New Evolutionary Synthesis Necessary?" *Science* 213: 967–71.

Tennyson, A. 1850. "In Memoriam." In *In Memoriam: An Authoritative Text, Backgrounds and Sources Criticism*, ed. R. H. Ross, pp. 3–90. New York: Norton.

Thompson, D. W. 1948. *On Growth and Form*. 2nd ed. Cambridge: Cambridge University Press.

Vogel, S. 1988. *Life's Devices: The Physical World of Animals and Plants*. Princeton, NJ: Princeton University Press.

Wagner, G. P. 1988. "The Influence of Variation and of Developmental Constraints on the Rate of Multivariate Phenotypic Evolution." *Journal of Evolutionary Biology* 1: 45–66.

Wallace, A. R. 1855. "On the Law Which Has Regulated the Introduction of New Species." *Annals and Magazine of Natural History* 16: 184–96.

———. 1858. "On the Tendency of Varieties to Depart Indefinitely from the Original Type." *Journal of the Proceedings of the Linnean Society, Zoology* 3: 53–62.

———. 1864. "The Origin of Human Races and the Antiquity of Man Deduced from the Theory of Natural Selection." *Journal of the Anthropological Society of London* 2: clvii–clxxxvii.

———. 1870. *Contributions to the Theory of Natural Selection: A Series of Essays*. London: Macmillan.

———. 1876. *The Geographical Distribution of Animals*. 2 vols. London: Macmillan.

———. 1889. *Darwinism: An Exposition of the Theory of Natural Selection with Some of Its Applications*. London: Macmillan.

———. 1900. *Studies: Scientific and Social*. London: Macmillan.

———. 1905. *My Life: A Record of Events and Opinions*. London: Chapman and Hall.

Waters, C. K., and A. van Helden, eds. 1992. *Julian Huxley: Biologist and Statesman of Science*. Houston: Rice University Press.

Weikart, R. 2004. *From Darwin to Hitler: Evolutionary Ethics, Eugenics, and Racism in Germany*. New York: Palgrave Macmillan.

Weldon, W. F. R. 1898. Presidential Address to the Zoological Section of the British Association. *Transactions of the British Association*, pp. 887–902. Bristol.

Whewell, W. 1837. *The History of the Inductive Sciences*. 3 vols. London: Parker.

———. 1840. *The Philosophy of the Inductive Sciences*. 2 vols. London: Parker.

Williams, G. C. 1966. *Adaptation and Natural Selection*. Princeton, NJ: Princeton University Press.

Wilson, C. 2006. "Kant and the Speculative Sciences of Origins." In *The Problem of Animal Generation in the 17th and 18th Centuries*, ed. J. Smith, pp. 375–401. Cambridge: Cambridge University Press.

Wilson, D. S. 2002. *Darwin's Cathedral*. Chicago: University of Chicago Press.

Wilson, E. O. 1978. *On Human Nature*. Cambridge, MA: Harvard University Press.

———. 1984. *Biophilia*. Cambridge, MA: Harvard University Press.

———. 1992. *The Diversity of Life*. Cambridge, MA: Harvard University Press.

———. 1994. *Naturalist*. Washington, DC: Island Books/Shearwater Books.

———. 2006. *The Creation: A Meeting of Science and Religion*. New York: Norton.

Wright, L. 1973. "Functions." *Philosophical Review* 82: 139–68.

Wright, S. 1931. "Evolution in Mendelian Populations." *Genetics* 16: 97–159.

———. 1932. "The Roles of Mutation, Inbreeding, Crossbreeding and Selection in Evolution." *Proceedings of the Sixth International Congress of Genetics* 1: 356–66.

INDEX

abortion, 217, 234

adaptations, 25, 39, 48, 63, 126, 143, 182–83, 205

    adaptation context, 59

    adaptive advantages, 192

    adaptive complexity, 21, 34, 182–83, 186

    adaptive landscapes, 107–19

    adaptiveness of organisms, 77, 186, 188

    advantageous adaptations, 138

    and change, 184

    and continuance, 45

    cooperation as adaptive strategy, 145

    to environmental challenge, 186, 188

    of fish, 67–68

    of God, 210

    and natural selection, 20–21, 59, 67, 68, 147, 183, 199, 224

    nonadaptation, 191

    of orchids, 182–83

    preventing adaptation, 190

    religion as adaptive, 202–204, 206–208

    shifting balance of, 108–109. *See also* Shifting Balance Theory

    See also change; contrivance; mechanism

aeoquivoca, generatio, 40

Agassiz, Louis, 104, 184, 185

agnosticism, 63, 65, 201, 216, 217

    Darwin as an agnostic, 34, 66, 77

American Museum of Natural History, 137

American Pragmatism, 114

Anglicanism, 74, 96, 209, 221, 226

animals, 40, 180, 184, 218, 221

    breeding of, 98–99

    characteristics reflecting what they do, 187–88, 199, 207–208

    mummified, 47, 220

    order in the animal world (*embranchements*), 46–47, 48

    similarities between, 23, 180

Antennapedia complex gene, 189–90

anti-Enlightenment, 209

"antimony," 37

archetypes, 43, 184, 189, 190

architectural constraints, 191

Aristotelianism, 54

Aristotle, 178, 179, 180, 195, 224

Arnold, Thomas, 124

artificial selection, 76, 82, 98–99, 224. *See also* natural selection

atheism, 34, 124, 216, 217, 219, 224, 235

    atheistic humanism, 216

    "new atheism," 217, 235

Atran, Scott, 207–208

Australopithecus afarensis, 64–65

Autobiography (Darwin), 200

Back to Methuselah (Shaw), 159

Bacon, Francis, 36, 55

balance, shifting. *See* Shifting Balance
  Theory
Balfour, Frank, 184
Barnacle books (Darwin), 59
barnacles, Darwin on, 18, 59, 64, 75
Bates, Henry Walter, 77, 79, 96, 225
Baupläne, 43, 189, 190
HMS *Beagle*, 18, 75, 76, 81. See also
  Voyage of the Beagle, The (Darwin)
Bergson, Henri, 113–14, 125–26, 132
biodiversity, 132, 133, 233
biological determinism, 169
blind-law explanation, 33, 41
Bowler, Peter, 96
Boyer, Pascal, 206–207
Boyle, Robert, 53–55, 56, 57, 62
Brave New World (Huxley), 238
*Breaking the Spell: Religion as a Nat-
  ural Phenomenon* (Dennett), 205
breeder, world of, 19, 98–99, 222
Brer Rabbit (fictional character), 215
Brewster, David, 27, 153
Bridgewater Treatises, 21, 74
British Association for the Advance-
  ment of Science, 74
Broad, C. D., 134–36, 137, 148
Buck (fictional character), 157–58
butterfly mimicry, 96, 225
by-products
  religion as, 202, 206–208
  of selection, 199

Call of the Wild, The (London), 157–
  58, 159, 170
Calvin, Jean, 204–205
Calvinism, 203, 209, 221
*Cambridge Companion to the "Origin of
  Species"* (Ruse and Richards), 17
Can a Darwinian Be a Christian
  (Ruse), 216

Capital (Marx), 103
Carnegie, Andrew, 228–29
Carrie (fictional character), 158–59
Castle, W. E., 97
Categorical Imperative, 132
Catholicism, 11, 174, 203, 208–209,
  217, 237
cause, 94, 96, 145, 193, 199, 208, 220,
  237
  causal connections, 44
  causal force, 57, 66
  causality, 38, 42, 100, 227
  causal theory of evolution, 36
  formal causes, 195
  God as cause of causes, 219
  immaterial cause, 43
  material causes, 53, 195
  true causes, 74, 76
  See also final causes; first cause
Chamberlain, Neville, 215
Chambers, Robert, 33, 81, 86, 154,
  221, 229
change, 48, 96, 108, 114, 124, 125,
  128, 138, 178, 186, 194, 205
  and adaption, 183, 184
  brought about by breeders, 19, 113
  cosmical change, 100, 227
  cycle of change, 29
  of equilibrium, 115
  evolutionary change, 47, 66, 67, 82,
    95, 110–11, 185
  and humans, 127, 131, 217–18
  mechanisms of, 15, 115–16, 222
  model of change, 19
  and natural selection, 57, 183, 199
  in nature, 19–20
  organic change, 49, 76, 96
  progressive change, 127, 138
  species change, 82, 96, 104
  See also adaptations; natural selection

Charles I (king), 209
Christianity, 62, 94, 201, 206, 208–
    209, 211–12, 217, 219–20, 228,
    229, 234
  belief in the soul, 25–26
  challenge for, 221
  and Darwin, 63, 77, 174–75, 201,
      216, 221, 224–25, 236
    and acceptance of Darwinian
        evolutionary theory, 216
    and acceptance of descent with
        modification, 25
  and Dawkins, 236–37
  vs. evolution, 226, 232
    evolution as alternate to, 102,
        105, 232, 236
    seeing Darwinism as evil, 170
  revitalization of, 230
  See also names of specific groups, i.e.,
      Anglicanism, Calvinism, *etc.*
Clarissa (fictional character), 160–62,
    166, 168
coiling of shells, 193–94
complexity, 20, 33, 99, 125, 133, 185,
    187
  adaptive complexity, 21, 34, 182–
      83, 186
  and evolution, 27, 100–101, 115,
      127, 139
  and humans, 166
  of machines, 60, 64, 126, 182
  of nature, 41, 60–61, 98
  organized complexity, 21
Comte, August, 27
conditions of existence, 45–47, 220.
    See also final cause
Conklin, E. G., 104–105
"consilience of inductions," 21–23,
    99, 223
constraints, 24–25, 185–86

architectural constraints, 191
developmental constraints, 186,
    188–89
genetic constraints, 187–88
against open coiling of shells, 193–
    94
physical constraints, 192–94
structural constraints, 191–92
contrivance, 20, 41, 57, 61, 182, 183,
    199
  mechanical contrivance, 51, 53, 55,
      56, 57, 58, 59, 67, 68, 69
  nature as contrivance, 62–65
  primitive contrivance, 54
  See also adaptations
control, 143, 156, 211
  and independence, 125, 131, 133–
      34, 136–37, 138, 139
  and mechanisms, 65
Cope, Edward D., 104
correlation of parts, 46
cosmical change, 100, 227
Coyne, Jerry, 215, 216
crab carapace dimensions, 96
creationism, 34, 173
  are creationists right?, 215–41
  claim that evolution is a religion,
      174–75
  neo-creationism, 21
  See also Intelligent Design Theories
creative evolution, 159
Creative Evolution (Bergson), 113, 125
Crick, Francis, 173
Critique of Judgement (Kant), 36, 39,
    42, 44
culture vs. genetics causing change,
    205
Cuvier, Georges, 33, 36, 44–49, 104,
    137, 220–21

Darwin, Charles Robert, 15
  as an agnostic, 34, 66, 77
  as author
    Autobiography, 200
    Barnacle books, 59
    Descent of Man, 84, 101, 200–201, 223
    Essay, 58
    Notebooks, 58
    On the Origin of Species by Means of Natural Selection, or the Preservation of the Favoured Races in the Struggle for Life, 15, 17–28, 57, 59–60, 65, 75, 77, 78, 95, 98, 110–11, 153, 155, 181, 194, 199, 222–25, 227
    On the Various Contrivances by which British and Foreign Orchids Are Fertilized by Insects, and on the Good Effects of Intercrossing, 181–83
    Sketch, 58
    Variation under Domestication, 58
    Voyage of the Beagle, The, 17, 58–59
    Zoonomia, 41, 43, 76
  on the back of the English ten-pound note, 18
  and Christianity, 63, 77, 174–75, 201, 216, 221, 224–25, 236
  and acceptance of Darwinian evolutionary theory, 216
  and acceptance of descent with modification, 25
  comparison with Freud, 26
  comparison with Spencer, 97–101
  comparison with Wallace, 78–83
  on concept of mechanism, 57–61
  discovery of natural selection, 75–77

  on religion, 63, 200–201, 210, 211, 224
Darwin, Erasmus, 15, 33, 39, 41, 76, 218–19, 221, 232
Darwinism, 99, 110, 121, 170, 185, 192, 225, 229, 236
  explaining religion, 199–213
  and mechanism, 51–72
  neo-Darwinism (synthetic theory of evolution), 69, 96, 124, 229–31
  See also Social Darwinism
Darwinism (Wallace), 78
Darwinismus movement, 184
"Darwin's Bulldog." *See* Huxley, Thomas Henry
Dasypodius, Cunradus, 53
Dawkins, Richard, 19, 93, 99, 169, 175, 186, 192, 205, 215, 233, 235–37
de Clérambault's syndrome, 160–61, 167, 168
deism, 34, 54, 63, 77, 212, 219, 221, 224
deism-cum-progress, 221
Dennett, Daniel, 205–206, 208–209, 210, 215, 238
Derrick (fictional character), 155–56
Descartes, René, 52, 55
descent
  from common ancestor, 40
  with modification, 17, 22–23, 24, 25, 67
  acceptance of by Christians, 25
  See also homologies
Descent of Man (Darwin), 84, 101, 200–201, 223
design, 76, 182, 191, 236
  arguments for concept of, 181
  divine design, 74, 77, 186, 224. *See also* creationism; Intelligent Design Theories

of final cause, 25, 35, 159. *See also*
   final causes
first design, 54
in nature, 37, 38–39, 49, 55
and organisms, 45, 46, 67, 75
re-design, 133
determinism, 168–69, 225
   biological determinism, 169
   evolutionary determinism, 216
   fatalistic determinism, 168
developmental constraints, 186, 188–
   89
Dickens, Charles, 18
Dicrocelium dendriticum. See lancet
   fluke
dignity, 140, 141
*Disquisition about the Final Causes of
   Natural Things* (Boyle), 55
divine design, 74, 77, 186, 224
   first design, 54
   See also creationism; Intelligent
      Design Theories
DNA, 187, 188, 208, 236
Dobzhansky, Theodosius, 66, 97,
   106, 110, 112–13, 185, 230
"Dover Beach" (Arnold), 125
Dreiser, Theodore, 158–59, 168
Drosophila, 188–90
Duncan, D., 104
dynamic equilibrium, 104, 110–19,
   132, 227

ecological genetics, 230
Ecological Genetics (Ford), 230
Edward VI (king), 209
Eldredge, Niles, 116
Elizabeth (queen), 209
embranchements of animals, 46–47, 48
embryology. *See* evolutionary
   development

embryos, similarity of, 223
Enduring Love (McEwan), 94, 160–62
Enlightenment, 124, 237
   anti-Enlightenment, 209
Entemann, Wilhelmine Marie (Key),
   114
epistemology, 9, 63
   epistemic excellence, 63–64, 69
   epistemic norms, 56
   epistemic virtues, 52, 64
equilibrium
   change of, 115
   dynamic equilibrium, 110–19, 132
   maintaining equilibrium. *See*
      heredity
   punctuated equilibrium theory, 24,
      116–17
   reflective equilibrium, 145
Essay (Darwin), 58
ethics
   and evolution, 121–51, 226
   metaethics, 128, 129–30, 141–48
   naturalistic ethics, 143, 144, 146
   natural nature of, 231
   new ethics, 94, 140
   nonethical nature of, 143
   normative ethics, 133, 141–42,
      144–45, 148
   substantive ethics, 147
   eugenics, 133
evo-devo. *See* evolutionary development
evolution
   after Darwin, 65–68
   and blind-law explanation, 41
   causal theory of evolution, 36
   vs. Christianity, 226, 232
      evolution as alternate to Chris-
         tianity, 102, 105, 232, 236
   comparison of Darwin and Spencer,
      97–101

and complexity, 27, 100–101, 115, 127, 139
control and independence as a factor, 136–37
creative evolution, 159
established by *On the Origin of Species by Means of Natural Selection* (Darwin), 15
and ethics, 121–51, 226
evolutionary change, 47, 66, 67, 82, 95, 110–11, 185
evolutionary humanism, 128–32, 211
evolutionism, 232
final cause as argument against, 48, 49
and humans, 122, 127, 131, 132, 145, 199, 221
  as pinnacle of evolution, 102, 125, 136, 138, 139, 154
Julian Huxley on, 134–37
Kant on, 39–41
linked to individual growth, 185
moral issues
  evolution justifying morality, 144
  promoting the ends of evolution, 132
and the novel, 94, 153–70
organic evolution, 35–50, 39, 251
and progress, 130, 133, 134, 226–28, 233
punctuated equilibrium theory, 24
as religion, 215–41
  as alternative to religion, 233
  creationist claim that evolution is a religion, 174–75
  Darwinism explaining religion, 199–213
Simpson on, 138
See also Darwinism; natural selection; neo-Darwinism (synthetic theory of evolution); *On the Origin of Species by Means of Natural Selection, or the Preservation of the Favoured Races in the Struggle for Life* (Darwin); Shifting Balance Theory; Social Darwinism
Evolution: The History of an Idea (Bowler), 96
Evolution: The Modern Synthesis (Huxley), 124, 130
Evolutionary Analysis (Freeman and Herron), 66–67
evolutionary biology, 67, 68, 69, 93–94, 159, 169, 177, 232, 235
  Spencer's contributions to, 95–119
evolutionary determinism, 216
evolutionary development, 173, 177–98
evolutionary divide
  form across the divide, 183–85
  function across the divide, 180–83
"Evolutionary ethics" (lecture by Julian Huxley), 130
evolutionary humanism, 128–32, 211
evolutionary naturalism, 146
Evolutionary Naturalism: Selected Essays (Ruse), 144
existence, 42, 43, 48, 100–101, 116, 208, 211, 221, 227, 236
  conditions of existence, 45–47, 220
  of a deity, 181, 201, 210, 224, 236–37
  struggle for, 10, 80–81, 84, 95, 98, 145, 159
extinction, 126–27

fatalistic determinism, 168
fiction and evolution, 94, 153–70

final causes, 25, 37, 38
  as argument against evolution, 48, 49
  and blind-law explanation, 41
  Cuvier on, 45–47
  design and, 25, 36
  doing science without, 55–56
  evolution going against, 40
  form and function, 179
  importance of material and formal
    causes, 195
  Kant on, 36–39, 44
  organisms exhibiting, 77
  as vestal virgins, 36, 55
  without design or a Designer, 39
first cause, 218
first design. *See* design
fish
  adaptations of, 67–68
  development of, 185
Fisher, Ronald A., 96, 105, 108, 110,
  111, 116, 185, 230
Fitness of the Environment, The (Henderson), 115
flexibility, 125, 127, 143
Fodor, Jerry, 57
Ford, E. B., 185, 230
form
  across the evolutionary divide, 183–
    85
  formalists, 190, 194
  and function, 178, 195
formal causes, 195
Foucault, Michel, 36, 49
Freeman, Scott, 66–67
Freud, Sigmund, 26, 130
From Darwin to Hitler (Weikart), 170
fruit flies, 188–90
function
  across the evolutionary divide, 180–
    83

dominant feature of living beings,
  185–86
form and function, 178–80, 195
functionalists, 189–90, 191, 193,
  194
functioning nature of organisms,
  220–21
selection producing things that
  function, 199
fundamentalism, 215. *See also*
  creationism

Galapago Archipelago, 22–23, 76
Gegenbaur, Carl, 184
Genesis Flood (Morris), 216
Genetical Theory of Natural Selection,
  The (Fisher), 96
genetics
  coming of Mendelian genetics, 229
  vs. culture causing change, 205
  ecological genetics, 230
  genetic constraints, 187–88
  genetic counseling vs. eugenics,
    133
  genetic drift, 110
  homeotic genes, 188
  Hox genes, 188, 189–90
  phylogenetic inertia, 190
  population genetics, 96, 230
  See also heredity
Genetics (Sewall), 106
Genetics and the Origin of Species
  (Dobzhansky), 66, 97, 230
genomes, large, 187
Gilbert, S. F., 177
God
  allowing humans to struggle, 225
  belief in, 201
  Boyle's view of, 53–55
  as cause of causes, 219

as designer, 77, 186. *See also* Intelligent Design Theories
God-people relationship, 205
proving the existence of, 37–38, 210–12, 224, 236–37
science proving and glorifying, 74–75
as an unmoved mover, 219
Will of God, 144
world as creation of, 52, 54, 65
See also creationism; divine design
God Delusion, The (Dawkins), 93, 215, 236–37
Goethe, Johann Wolfgang von, 43, 183, 184
Goldberg, Rube, 61
Gould, Stephen Jay, 21, 24–25, 116–17, 169, 189, 190, 191, 192, 206
Grant, Peter, 21
Grant, Robert, 76
Grant, Rosemary, 21
Gray, Asa, 62, 77, 224
Greene, John C., 121, 131
Greene, Toby (fictional character), 160–61
group selective analysis, 204–205

Haeckel, Ernst, 184, 185
Haldane, J. B. S., 96, 105, 185
Hallam, Arthur, 154–55, 221–22
Hamilton, William, 83, 233–34, 238
harmonious correlation, 105
Harré, Horace Romano, 55, 56
Harvard University, 97, 104, 114, 132, 137, 160, 224, 232
Hegel, Georg Wilhelm Friedrich, 43
Henderson, L. J., 114–15
Henry VIII (king), 209
Henslow, John, 75
Heraclitus, 15

heredity, 24, 96, 116, 124, 140, 229. *See also* genetics
Herron, Jon C., 66–67
Herschel, John F. W., 62, 74, 75, 76, 200
heterogeneity, 93, 100, 104, 110–11, 227
heteronyma, generatio, 40
Hitler, Adolf, 170, 215
homeotic genes, 188
homogeneity, 111, 132, 227
homologies, 23, 104, 179, 180, 189, 190
Unity of Type, 195
homonyma, generatio, 40
homosexuality, 26, 217, 234
Houseman, A. E., 236
Hox genes, 188, 189–90
humanism
atheistic humanism, 216
evolutionary humanism, 128–32, 211
secular humanism, 175, 206
humans, 86, 94, 99, 102, 165, 181, 207, 216–17, 218, 219
complexity of, 166
and evolution, 122, 127, 131, 132, 145, 199, 221
as pinnacle of evolution, 102, 125, 136, 138, 139, 154
mummified, 47
non-specialization of, 139, 230–31
origins of, 83, 84
similarity to other organisms, 177, 189, 223
Hume, David, 122, 133, 147, 148, 169, 174, 200, 217
Huxley, Aldous, 124
Huxley, Julian, 66, 94, 123–37, 175, 231

and Simpson, 138, 141, 142, 147–48

Huxley, Thomas Henry, 34, 65–66, 75, 122, 124, 146–47, 148, 184, 225–26, 229, 231, 232
and Simpson, 139
Hyatt, Alpheus, 184

If I Were Dictator (Huxley), 129
"illusion of the genes," 146
independence, 125, 133–34, 136, 139
individual growth, evolution linked to, 185
individual responsibility, 144, 148
industrial progress, 218
In Memorium (Tennyson), 154–55, 159, 221
integrity, 140
intelligence, 125
Intelligent Design Theories, 9–10, 21, 34, 234
See also creationism
Iolanthe (Gilbert and Sullivan), 195
Islam, 206, 212, 217
isomorphisms of organisms, 39–40, 193–94

Jesus, 217, 229, 234
"Jesus number," 192–93
Johnson, Phillip, 234–35
Journal of Researches into the Geology and Natural History of the Various Countries Visited by HMS Beagle (Darwin). *See Voyage of the Beagle, The* (Darwin)
Judaism, 203, 204, 206, 212

Kant, Immanuel, 33, 34, 35–50
Kettlewell, H. B. D., 21
Keynsian economics, 141

Kim (Kipling), 102–103
Kings College, 124
kin selection, 233
Kipling, Rudyard, 102–103
K-selectionist, 203–204
Kuhn, Thomas, 9, 178, 179, 225

Lacey, Joseph (fictional character), 160–61
laissez-faire society, 102, 131, 227–28, 229
Lamarck, Jean Baptiste de, 33, 76, 220, 221
Lamarckism, 26, 95–96, 101, 111, 184, 219
lancet fluke, 205–206
LEGOs, 25
L'Évolution Créatice (Bergson). *See Creative Evolution* (Bergson)
Lewontin, Richard, 169
literature and evolution, 153–70
Logan, John (fictional character), 160, 161, 162–63, 167–68, 169
London, Jack, 157, 168
London Zoo, 124
"Lucy" (hominid), 64–65
Lutheranism, 209
Lyell, Charles, 62, 76, 154
Lysenko, Trofim, 141, 231

Mackintosh, James, 200
macromutations, 66
Major Features of Evolution. See Tempo and Mode in Evolution (Simpson)
Malthus, Thomas Robert "Robert," 19, 80, 224–25
"Man the Philosophizer" (Huxley), 125
Martin, H. N., 184
Marx, Karl, 103

Mary (queen), 209
material causes, 53, 195
materialism
    mechanical materialism, 43
    scientific materialism, 211, 233
Maynard Smith, John, 21, 186, 193,
    194
Mayr, Ernst, 93, 97, 106, 112, 230
McEwan, Ian, 94, 160–69
Meaning of Evolution, The (Simpson),
    139, 230
mechanism
    and control, 65, 128
    and Darwinism, 51–72
    mechanical contrivance, 51, 53, 55,
        56, 57, 58, 59, 67, 68, 69
    mechanisms of change, 15, 115–16,
        222
Mein Kampf (Hitler), 105, 170
meme theory, 205, 208, 216
Mendel, Gregor, 24, 96, 106, 124,
    229
metaethics, 128, 129–30, 141–48
methodological naturalism, 56
Mill, John Stuart, 85
Modern Synthesis (Huxley). *See Evolu-
    tion: The Modern Synthesis*
    (Huxley)
Molière, 80
Moore, Benjamin, 114
Moore, G. E., 121, 122, 134, 148
moral issues, 9, 226
    evolution justifying morality, 144
    fighting Intelligent Design as a
        moral crusade, 10
    Hume's moral theory, 147
    Julian Huxley on, 128–29, 132–33
    matters of facts vs. morality, 133–
        36, 148
    moral responsibility, 169

normative morality, 146
not always religious, 235
objective existence of morality,
    145–46
obligation to promote progress,
    227–28
promoting the ends of evolution,
    132
relativity of morality, 128
and religion, 204
Mormonism, 208
morphology, brain, 187
Morris, Henry, 216
Muslims. *See* Islam
mutations, 140, 188
    macromutations, 66
    transmutationism, 17, 41, 101, 102,
        218

Narrow Roads of Gene Land, The
    (Hamilton), 234
National Socialist Philosophy, 170
Natural History of Religion (Hume),
    174, 200
naturalism, 144
    evolutionary naturalism, 146
    methodological naturalism, 56
    naturalistic accounts of religion,
        174, 210, 211, 212
    naturalistic ethics, 143, 144, 146
    naturalistic fallacy, 121, 133, 144,
        146, 148
    naturalistic theory, 57, 84
    scientific naturalism, 211, 233
natural selection, 18–21, 183, 222–23,
    227, 236
    and adaptations, 20–21, 59, 67, 68,
        147, 183, 224
    and biological trends, 143
    by-products of selection, 199

and change, 57, 183, 199

Darwin on, 98–99

and evolution

as dominant mechanism of evolutionary change, 185–86, 229

as key to evolution, 199

as mechanism for evolution, 27, 63–66, 212

new role in evolutionary development, 177–78

explaining homologies, 23

gaining acceptance of theory, 95–96

and intelligence, 84

Julian Huxley on, 127

metaphor of, 58

producing ethical sense, 231

and progress, 99–100, 113–14

and religion, 20, 202

research needed, 173

selection producing things that function, 199

Thomas Huxley on, 226

Wallace on, 73–89

See also adaptations; artificial selection; change; Darwinism; sexual selection; Social Darwinism; "survival of the fittest"

Natural Theology (Paley), 20, 63, 224

nature

change in nature, 19–20

complexity of, 41, 60–61, 98

as contrivance, 62–65

Darwin's perception of, 58

design in nature, 37, 38–39, 49, 55

as an entity, 53

functioning nature of organisms, 220–21

Kant's philosophy of nature, 35

key to understanding of, 220

as a machine, 51–72

nature of after Darwin, 57

as a product of design, 181

"red in tooth and claw," 154, 168

as a system of stages, 43

world of nature, 222

Naturphilosophen, 57, 104, 184

"nebular hypothesis," 35

neo-creationism. *See* Intelligent Design Theories

neo-Darwinism (synthetic theory of evolution), 69, 96, 124, 229–31

neo-Paleyism, 186

"new atheism," 217, 235

"New bottles for new wine" (Huxley), 131

new ethics, 94, 140

Newton, Isaac, 53, 74

nihilism, 36, 235–36

Noah and the flood, 221

normative ethics, 133, 141–42, 144–45, 148

normative morality, 146

Norris, Frank, 155, 168

Notebooks (Darwin), 58

novel and evolution, 94, 153–70

Nuremberg Trials, 168

objectivists, 56

objectivity, 9, 39, 52, 64, 65, 69, 136, 145, 164–65, 196

objective reality, 9, 37, 51

progress as objective fact, 127, 136, 139, 148, 231

Octopus, The (Norris), 155, 159–60

Oken, Lorenz, 184

On Human Nature (Wilson), 164, 233

On the Origin of Species by Means of Natural Selection, or the Preservation of the Favoured Races in the Struggle for Life (Darwin), 15, 17–

28, 57, 59–60, 65, 75, 98, 153, 155, 199, 222–25
comparison with Shifting Balance Theory, 110–11
publishing of, 17, 18, 77, 78, 95, 181, 194, 222, 227
purpose of, 95
On the Various Contrivances by which British and Foreign Orchids Are Fertilized by Insects, and on the Good Effects of Intercrossing (Darwin), 181–83
ontogeny, 185
Opitz, J. M., 177
Opus Postumum (Kant), 41, 42–43
orchids, Darwin on, 60–61, 63, 64, 75, 181–83, 188, 225
organic change, 49, 76, 96
organic evolution, 35–50, 154, 221
organismic analogy, 115
organisms, 67, 80–81, 189
    adaptiveness of, 77, 186, 188
    balanced organisms, 132
    control and independence in, 125, 131, 133, 133–34, 136–37, 138, 139, 230–31
    Cuvier on, 45–47, 48
    and design, 45, 46, 67, 75
    evolution linked to individual growth, 185
    functioning nature of, 220–21
    humans similarity to other organisms, 177, 189, 223
    isomorphisms of, 39–40
    Kant on, 36–39, 45
    nature of after Darwin, 57
    as one archetype, 184
    ontogeny recapitulates phylogeny, 185
    shared genes, 177

and Shifting Balance Theory, 107
    variations in, 222
organized complexity, 21
Origin of Species. See On the Origin of Species by Means of Natural Selection, or the Preservation of the Favoured Races in the Struggle for Life (Darwin)
Orwell, George, 215
Owen, Richard, 18, 23, 183–84

Paley, William, 20, 49, 57, 63, 77, 182, 186, 224
panpsychic monism, 112
paradigm, use of the term, 177–79, 194–95, 196, 225
Parmenides, 15
Parry, Jed (fictional character), 160–61, 164, 165, 166
Paul, Saint, 168–69, 229, 235
personal responsibility, 140, 146, 231
phylogeny, 185
    phylogenetic inertia, 190
physical constraints, 192–94
physical reality, 109
Plato, 57, 184
poetry, use of for evolutionary ideas, 154
Popper, Karl, 9, 64
Popperians, 85
population genetics, 96
    theoretical population genetics, 230
Powell, Baden, 74
Presley (fictional character), 155–56
primitive contrivance, 54
Principia Ethica (Moore), 121, 122, 123
Principles of Biology (Spencer), 65
Principles of Geology (Lyell), 76
progress, 116, 124–25, 127–28, 168, 226

deism-cum-progress, 221
and evolution, 130, 133, 134, 226–28, 233
as fact of nature, 148
ideology of, 217–18
industrial progress, 218
maximizing control and independence, 131
and natural selection, 99–100, 113–14
obligation to promote progress, 227–28
progress as objective fact, 127, 136, 139, 148, 231
progressionism, 111–13, 117, 126, 131–32, 138–39
progressive change, 127, 138
social progress, 125, 155–56, 218
Protestant churches, 217, 220
Provine, Will, 107
psychological picture, 179
punctuated equilibrium theory, 24, 116–17

Quarterly Review (magazine), 18–19

Raff, Rudolf A., 177, 187, 189
Rawlsian theory of justice, 145
Ray, John, 180
reality
objective reality, 9, 37, 51
physical reality, 109
reduction analogy, 136
reflective equilibrium, 145
religion
as adaptive, 202–204, 206–208
Charles Darwin and religion, 224
Darwinism explaining religion, 199–213
creationist claim that evolution is a religion, 174–75
evolution as alternate to, 233
evolution as religion, 215–41
group selective analysis of, 204–205
human need for, 209–10
naturalistic accounts of, 174, 210, 212
as a natural phenomenon, 201
religion as by-product, 206–208
and reproductive rates, 203–204
search for knowledge as alternative, 141
secular vs. spiritual religion, 226
social implications of, 102
See also Christianity; deism; Islam; Judaism
"Religion without Revelation" (Huxley), 130
reproductive rates and religions, 203–204
responsibility, 129, 141, 142, 233
individual responsibility, 144, 148
moral responsibility, 169
personal responsibility, 140, 146, 231
reverse engineering, 63
Reynolds, Vernon, 203, 204
Reznick, David, 67–68
Rice Institute (now Rice University), 124
Richards, Robert J., 17, 43
Robinson, Heath, 61
Rockefeller, John D., 228
Rose, Joe (fictional character), 160–69
r-selectionist, 203–204
Russett, C. E., 104

Saint-Hilaire, Étienne Geoffroy, 183
saltationism, 66
SBT. *See* Shifting Balance Theory
Schelling, Friedrich Wilhelm Joseph, 43, 184

science
    nature of, 9, 51–52
    and the running of an efficient
        state, 129
scientific materialism. *See* materialism
scientific naturalism. *See* naturalism
Scientology, 209
secular humanism, 175, 206
Sedgwick, Adam, 75, 81, 221
selection. *See* artificial selection; nat-
    ural selection
selectionism, 113
Selfish Gene, The (Dawkins), 19, 169
sexual selection, 21, 191, 202
    and intelligence, 84
Shaw, George Bernard, 159, 168
Shelgrim (fictional character), 156,
    157
shells, coiling patterns of, 193–94
Shifting Balance Theory, 93, 106–
    109, 173, 215
    influence of Spencer on, 109, 110–
        16
"Shifting Balance Theory of Evolu-
    tion" (Castle), 97
Sidgwick, Henry, 122
simplicity. *See* homogeneity
Simpson, George Gaylord, 94, 112,
    123, 126, 134, 137–48, 230–31,
    232
    and Julian Huxley, 138, 141, 142,
        147–48
Sister Carrie (Dreiser), 158–59
size related to weight, 192–93
Sketch (Darwin), 58
Social Darwinism, 101–106, 122, 228
    The Call of the Wild as an example
        of, 157–58
social progress, 125, 155–56, 218
sociobiology, 202, 204, 208

specialization, 15, 138, 139, 230
    human specialization, 127, 231
    leading to extinction, 126–27
Spencer, Herbert, 17, 93, 94, 131–32,
    148, 174, 226–29, 232
    contributions to American evolu-
        tionary biology, 95–119, 147,
        233
    critique of, 121–22, 133, 134
    impact on novels, 15, 155, 157
    influence on Shifting Balance
        Theory, 109, 110–16
    and "survival of the fittest," 20, 65
    See also Synthetic Philosophy
Spencerian philosophy. *See* Synthetic
    Philosophy
spirit forces, 34
spiritualism, 82, 84–85, 86
Stebbins, G. Ledyard, 106, 112, 230
structural constraints, 191–92
Structure of Evolutionary Theory, The
    (Gould), 117
Structure of Scientific Revolutions
    (Kuhn), 178
subjectivity, 9, 52, 142, 144, 145, 196
subordination of characters, 46–47
substantive ethics, 147
"survival of the fittest," 20, 65, 101,
    229. *See also* natural selection
Synthetic Philosophy, 102, 104, 114,
    132, 155, 156
"synthetic theory of evolution." *See*
    neo-Darwinism
Systemics and the Origin of Species
    (Mayr), 97, 230

*Taking Darwin Seriously: A Naturalistic
    Approach to Philosophy* (Ruse), 144
Tanner, Ralph, 203, 204
teleology, 37–38, 41, 45, 47, 55, 179

Tempo and Mode in Evolution
  (Simpson), 112, 137, 230
Tennessee Valley Authority, 129, 131
Tennyson, Alfred, 154, 168, 221
Thompson, D'Arcy, 195
Tolstoy, Leo, 159–60
transcendentalism, 184
transmutationism, 17, 41, 101, 102,
  218
Tree of Life, 20, 222
TVA. *See* Tennessee Valley Authority

UNESCO, 124, 129, 131, 142
uniformitarianism, 154–55, 221
Unity of Type, 195
US Department of Agriculture, 111,
  113
Ussher, James, 25

Variation and Evolution in Plants (Stebbins), 106, 230
Variation under Domestication
  (Darwin), 58
vestal virgins, final causes as, 36, 55
*Vestiges of the Natural History of Cre-
  ation* (Chambers), 17, 33, 81–82,
  87, 154, 221
Vestiginarianism, 17
vitalism, 87–88, 113, 126
Voyage of the Beagle, The (Darwin), 17,
  58–59

Wagner, Gunter, 194
walking on water, 192–93

Wallace, Alfred Russel, 20, 34, 73–89
War and Peace (Tolstoy), 159–60
Watson, James D., 173
Wedgwood, Josiah, 76
weight related to size, 192–93
Weldon, Raphael, 96
Whewell, William, 22, 62, 74–75, 81,
  86, 99, 200, 223
Whitman, Charles Otis, 114
Williams, George C., 67, 83
Wilson, David Sloan, 204–205, 209
Wilson, Edward O., 147, 162, 164,
  167, 175, 202–203, 210–11, 232–
  33, 238
  and Spencer, 117, 132, 233
  See also Categorical Imperative
*Wisdom of God, Manifested in the Words
  of Creation* (Ray), 180
world
  as creation of God, 52, 54, 65
  as a machine, 51–72
Wright, Larry, 44–45
Wright, Quincy, 114–15
Wright, Sewall, 93, 106, 114, 138,
  173, 185, 215, 230
  influence on Dobzhansky, 66, 105,
  112–13
  population genetics, 96, 97
  See also Shifting Balance Theory

Zoonomia (Darwin), 41, 43, 76